THE
LUMBER
INDUSTRY
IN EARLY
MODERN
JAPAN

釣木之圖

Kiso loggers lower timbers from a precipice. Man at upper right, using a tree for a friction grip, very carefully lowers timber, one piece at a time, from the pile to his right. His companion soaks the rope as it slips around the tree to keep it pliable and cool. Two other men with pole hooks *(tobiguchi)* observe and guide the timber's descent. Reproduced from Tokoro Mitsuo, *The Wood-Cutting and Transporting System of Kiso* (Tokyo: Tokugawa Institute for the History of Forestry, 1977), scroll 1, plate 11. Courtesy of the Tokugawa Institute and its director, Ōishi Shinzaburō. This painting was originally produced during the 1850s as part of a two-scroll set by Tomita Ayahiko (1811–1877), a petty functionary in the *bakufu* intendancy at Takayama in Hida Province. The forty-two paintings in the set illustrate the process of Kiso lumbering from initial felling to transport of the wood to Edo.

THE
LUMBER
INDUSTRY
IN EARLY
MODERN
JAPAN

Conrad Totman

 UNIVERSITY OF HAWAI'I PRESS / HONOLULU

Printed in the United States of America

00 99 98 97 96 95 5 4 3 2 1

Library of Congress Cataloging-in-Publication Data
Totman, Conrad D.
The lumber industry in early modern Japan / Conrad Totman.
   p.   cm.
Includes bibliographical references and index.
ISBN 0–8248–1665–X (alk. paper)
   1. Lumber trade—Japan—History.   I. Title.
HD9766.J32T68   1995
338.4'7674'00952—dc20   94–38136
                                         CIP

University of Hawai'i Press books are printed
on acid-free paper and meet the guidelines for
permanence and durability of the Council on
Library Resources.

Designed by Kenneth Miyamoto

For Lee, Barbette, and Gail

CONTENTS

ILLUSTRATIONS

viii

PREFACE

IN EARLY modern Japan (c. 1570–1870), as elsewhere, the basic material triad of human life was food, clothing, and shelter. Japanese scholars have produced fine studies of that society's provisioning systems for all three categories of goods. In English, as chapter 1 indicates, several studies examine the systems for producing and purveying food and clothing. However, the arrangements that provided materials for shelter have not received book-length attention. This small volume attempts to address that deficiency.

The study speaks primarily to two groups of scholar-teachers: those specializing in Japan's economic history or early modern (Tokugawa) history and those whose specialty is the commercial, industrial, or forest history of some other society and who wish to use the Tokugawa case for comparative purposes. The first chapter reviews the English-language literature on Tokugawa commercial history, identifying published works as completely as possible. It also contains a brief interpretive narrative of that history to provide a context for the subsequent chapters, which focus closely on practice and problems in the commercial lumber industry. The book will have achieved its central purposes if, first, it enriches the understanding that Japan specialists have of Tokugawa commercial history by introducing them to another major industry of the day and provides their students with a useful bibliographical tool. Second, I hope that historians of other societies will find the work a helpful guide to accessible works on Tokugawa commerce

and also a reliable and instructive source of information and interpretation on the lumber industry.

The gathering of source materials for this study occurred
mainly during a year's stay in Tokyo. The Japan Foundation
funded that stay and Professor Ōishi Shinzaburō, director of the
Tokugawa Rinseishi Kenkyūjo (Tokugawa Institute for the History of Forestry) enabled me to use the institute's resources. I have
also benefited from the National Diet Library's splendid journal
collection, the resources of Yale University Library, and the library
of the Kyoto Center for Japanese Studies.

I thank the Forest History Society, Inc., for permitting me
to reprint from the *Journal of Forest History* the articles "Logging the Unloggable: Timber Transport in Early Modern Japan"
27 (October 1983): 180–191 and "Lumber Provisioning in Early
Modern Japan, 1580–1850" 31 (April 1987): 56–70, which
appear here somewhat revised as chapters 3 and 2 respectively. I
thank Yale's School of Forestry and Environmental Studies for
permitting me to reproduce several large photographs of Meiji-
period forestry and lumbering and the Tokugawa Rinseishi Kenkyūjo for permitting the reproduction of illustrations in the bilin-
gual volume edited by Tokoro Mitsuo, *The Wood-Cutting and
Transporting System of Kiso-Style* [Kisoshiki batsuboku unzai
zue] (Tokyo: Tokugawa Rinseishi Kenkyūjo, 1977), which repro-
duces a handsome pair of picture scrolls of Edo-period logging in
Kiso.

Yale University graciously granted me a leave of absence to
spend the year 1992–1993 teaching at the Kyoto Center for Japanese Studies, during which time I delightedly bicycled through the
Yamaguni and Kitayama forest districts and began preparing this
manuscript for publication. I am grateful to Susan Hochgraf for
preparing the maps and the illustrations taken from sources other
than Tokoro's book. And finally, I thank the editors and reviewers
of the University of Hawai'i Press for catching errors and providing helpful suggestions and guidance.

Map 1. **Japan.** Based on *Nihon rekishi chizu* (map volume of *Nihon rekishi daijiten* [Tokyo: Kawade shobō shinsha, 1961]), map no. 33.

Note: In the early Tokugawa period the northeast was divided into only two provinces, rather than the seven shown here. Mutsu included nos. 1, 2, 4, 6, and 7; nos. 3 and 5 were known as Dewa.

0	100	200	300 Km.

Map 2. **Provinces of Japan.** Reproduced from Conrad Totman, *The Green Archipelago: Forestry in Preindustrial Japan* (Berkeley and Los Angeles: University of California Press, 1988), map no. 4. Courtesy of the University of California Press.

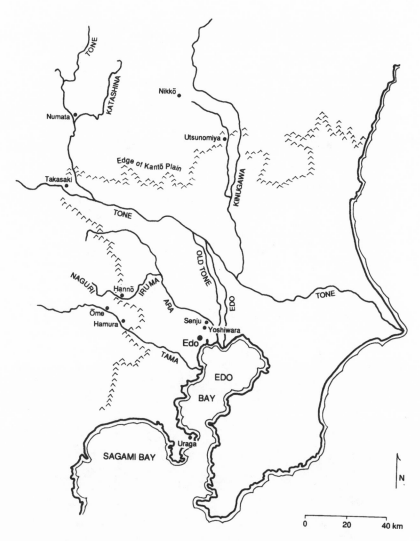

Map 3. **Rivers of the Kantō Plain.** Adapted from *Shōwa Nihon chizu*
(Tokyo: Tokyo Kaiseikan, 1941), map no. 15.

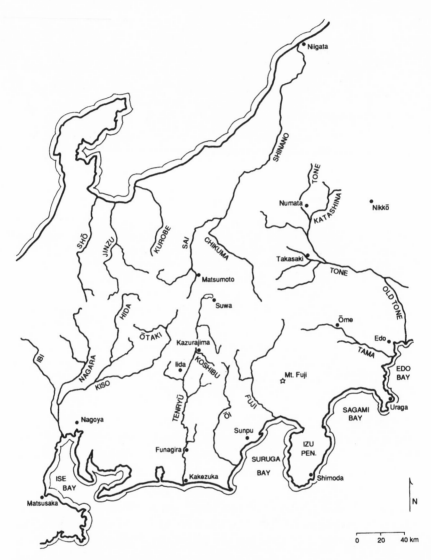

Map 4. **Major Rivers of the Chūbu Region.** Based on *Shōwa Nihon chizu* (Tokyo: Tokyo Kaiseikan, 1941), map no. 22.

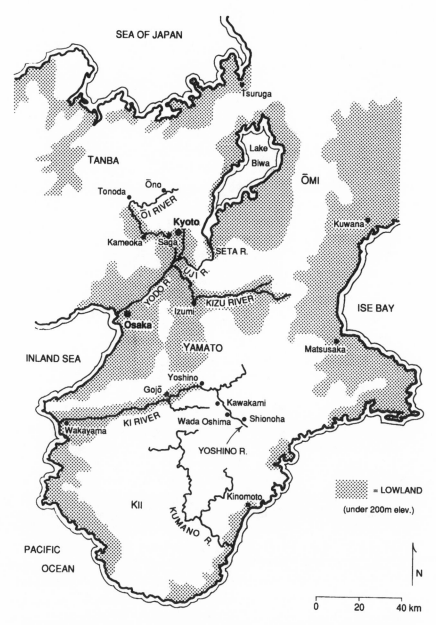

SEA OF JAPAN

Tsuruga

TANBA

Lake
Biwa

ŌMI

Ōno
Tonoda
ŌI RIVER
Kyoto
Kameoka Saga
SETA R.
UJI R.
KIZU RIVER
Izumi
YODO R.
Osaka

INLAND SEA

YAMATO

Kuwana

ISE BAY

Matsusaka

Yoshino
Gojō
Kawakami
KI RIVER Wada Oshima Shionoha
Wakayama
YOSHINO R.

KII

Kinomoto

KUMANO R.

PACIFIC

OCEAN

N

= LOWLAND

(under 200m elev.)

0   20   40 km

Map 5. **The Kinki Region.** Adapted from *Shōwa Nihon chizu* (Tokyo: Tokyo Kaiseikan, 1941), map no. 29.

INTRODUCTION

WE COMMONLY envisage human history as a series of stages in which hunter-gatherer forms of survival have given way to agricultural patterns of sustenance and, in recent generations, to an urban industrial order. Setting aside the provocative question of whether this overall process is more wisely celebrated or lamented, one may suggest that it is better viewed as a cumulative process rather than a series of stages.

Hunter-gatherers approached the biosystem as autonomous actors—"top predators"—who survived by exploiting whatever "natural" or "wild" creatures seemed useful. Agriculturalists domesticated much of their food and clothing supply by "taming" certain animals and replacing areas of natural growth with cultivated crops. In essence they employed interspecies alliances to form "homocentric biological communities" in which the humans flourished by collaborating with select plants and animals, such as maize, rice, yams, goats, pigs, and horses. This arrangement enabled the favored species to multiply and extend their dominion over more and more terrain, at the expense of other flora and fauna, in exchange for satisfying human wants. Despite the mutual advantages of this communal order, however, agriculturalists continued to depend on the surviving untamed sources for many necessities and amenities. Today, urban industrial society relies on both agricultural and undomesticated sources of supply for the materials that its members ceaselessly fabricate into usable goods.

As instructive as this triune categorization of hunter-gatherer, agriculturist, and industrial order may be, however, it is also useful to distinguish between preindustrial society as a whole, on the one hand, and industrial society on the other. What distinguishes the two is that the former survived on current biological production while the latter has added to its repertoire of exploitable natural resources the accumulated residue of ancient biomass—coal, oil, natural gas—which producers assiduously locate, extract, and process for myriad human uses. Because of the crucial role this ancient biomass now plays in human survival and because of the amount of labor involved in its extraction, processing, and purveying, it is tempting to characterize today's society as essentially a hunter-gatherer system whose level of exploitative intensity dwarfs that of our ancestors. Our dependence on those among us who actually handle these provisioning tasks is arguably little diminished from days of cave dwellings, and it is appreciably greater than that of agricultural society, which derived so much of its limited array of goods from arable land and domesticated animals.

We denizens of industrial society are aware of the agriculturalists in our midst because they define themselves as such and are organized as vocal interest groups and because we still nurture romanticized images of them: the family farm, *furusato,* hardy rancher, simple peasant. We are much less aware of the providers of undomesticated materials; even rhetoric about "the hunt" and "the noble savage" sounds archaic to our ear. We forget the basic character of this class of people in part because their crafts have been more completely absorbed into industrial processes and their presence concealed under other rubrics: fisherman, lobsterman, logger, trapper, miner, oilman, gas man.

We also tend to disregard the hunter-gatherers in our midst because we recognize that homocentric biological communities are superior instruments for producing targeted goods. Accordingly we transform hunting-gathering into agricultural production wherever possible, extending techniques of domestication into ever more areas of natural habitat. Thus plantation culture is now widely practiced in the production of coffee, rubber, other tropical forest goods, and temperate-zone wood products. Similarly, farming techniques are gaining importance in seafood culture as well as in such diverse areas as pearl and pelt production. Ultimately, it appears, we aspire to farm the entire planet and, beyond that, to transform it into the site of a human monoculture.

Despite the extension of manipulative techniques farther and farther into the biosphere, however, at present industrial society still depends on the production of hunter-gatherers. Without their intensive exploitation of "natural" resources, both living and fossilized, today's industrial society could not survive.

Even during the agricultural era, when products of arable land were central to everyday life, society still depended on the yield culled from woodland and waste, a dependence clearly visible in the history of early modern (Edo-period or Tokugawa) Japan. Most strikingly, as a widely urbanized society that eventually numbered some three hundred cities and towns, the Tokugawa populace required a continual supply of construction and fuel materials. Indeed, the need was so great that in the later Edo period, the Japanese began "farming" woodland by replacing exploitative forestry with plantation silviculture.

That dependence notwithstanding, rulers and town dwellers displayed little recognition of their need for "wild" goods even though they acknowledged their reliance on the yield from agronomy. That orientation is suggested by the tax system, which was officially anchored in the output of arable land, and by the formal status system, which recognized only one nonurban group, the agriculturalist or *nōmin*. It was also evinced by economic writings of the day, which gave most of their attention to urban mercantile affairs and wet-rice agriculture, less to other field crops, and very little to the yield of forest and sea.

Modern scholarship has tended to share this Tokugawa inclination, dwelling most heavily on the agricultural and urban commercial aspects of economic life. English-language studies of the Tokugawa economy, although few compared to the corpus in Japanese, reflect this emphasis. In terms of the basic triad of food, clothing, and shelter, the first two, which were mainly agricultural products, have received scholarly attention whereas the third, mostly provided by hunting-gathering techniques, has not. In recent years, however, more Japanese scholarship has examined the exploitation of undomesticated resources, and that work makes it more and more possible to redress the balance in the English-language literature.

Except for sunlight, air, water, and soil, the undomesticated resources most important to Tokugawa society were forest growth, waterborne biota, and mineral wealth. The corresponding industries were woodcutting, fishing, and mining. A few essays in English have looked at fisherfolk and miners, and this slender volume examines woodsmen. Readers should keep in mind, however, that it focuses on a single aspect of forest work—providing lumber for urban use. Firewood, charcoal, and bamboo purveying, and even subsistence lumber production, are touched only marginally even though they were major activities in their own right. And the myriad forms of woodland food production, both vegetable and animal, are completely disregarded.

Still, this commercial lumbering was an industry of such magnitude and fundamental importance to Tokugawa society that it seems to merit at least this level of attention. It presents the student with particularly interesting problems because timber provisioning, which is a difficult and dangerous task under any circumstances, was profoundly so in this case thanks to Japan's steeply mountainous topography. One wonders how the physical obstacles to provisioning were overcome so that Japan's wooden cities could be maintained. Questions about industrial organization, financing, and the distribution of benefits demand attention, as do the mechanics of timber felling, transporting, and processing. In closing, the unstated Euro-American comparative context requires that a few words be said about the nonuse of heavy-duty lumber wagons, crosscut saws, and water-powered sawmills and about the enduring preeminence of wooden construction despite widespread timber scarcity in the later Edo period.

The particular characteristics of this lumber industry were determined by the goods it provided and the context within which it operated, but the industry also had many qualities in common with Edo-period industry more generally. Consequently it provides an apt example of Tokugawa commercial organization and practice as a whole and of the way those changed as time passed. In its role as exemplar of Tokugawa commerce, it provides an opportunity for us to reflect on how scholars have dealt with the Tokugawa economy, to consider how that commercial experience can fruitfully be placed in broader historical context, and in the process to identify the pertinent works in English.

1 | ENGLISH-LANGUAGE APPROACHES TO TOKUGAWA COMMERCE

WE HUMANS have engaged in commerce—the exchange of goods and services—since prehistoric times, and the invention and elaboration of monetary instruments have expedited such exchange, making it central to our history throughout the world. How commerce has worked, to whose advantage, and with what broader consequences have therefore become questions of considerable scholarly interest. Unsurprisingly, that interest has generated a diversity of interpretations and some lively and enlightening debate, a situation evident even in the modest English-language corpus on the Tokugawa economy.[1]

Historiography of the Tokugawa Economy

As with any history, the study of Japan's early modern economy has always been shaped by scholarly premises and purposes. During the Edo period itself, Confucian dogma held that commercial activity was improper for a gentleman and unworthy of scholarly attention, and much commentary was devoted to affirming or refuting that thesis.[2] The interpretation favored by scholarly apologists for the Tokugawa order envisaged society as properly led by an enlightened, socially responsible ruling caste of gentlemen-samurai, whereas the vulgar men of commerce were seen as constituting a potential threat to social order and virtue and to the samurai role as defender and exemplar of that order and virtue. This juxtaposition of ruler and merchant retained enough force in

1

early-twentieth-century thought so that entrepreneurs continued laboring to legitimize their activities by claiming that their actions sprang not from a base craving for wealth but from high motives of public service and commitment to social uplift.[3]

The dyadic formulation of rulers versus merchants found a more recent scholarly voice in a rich literature of more-or-less Marxian character that defined the Tokugawa rulers as "feudal" and Tokugawa merchants as "bourgeois." In this construction the central tension of Tokugawa history pitted these merchants against the feudal rulers, who ruthlessly exploited the masses in their futile struggle to prevent the rise of a merchant-dominated bourgeois society.[4] This formulation derived much strength from the voluminous evidence of public hardship, especially in the later Edo period, as well as from the status-conscious phrasing of much legislation and the agrarianist rhetoric of Tokugawa Confucian ideologues, whose attacks on merchants could easily be construed as representing government policy. By disregarding the myriad ways in which rulers actually collaborated with merchants, thereby channeling commercial wealth into government coffers in return for diverse boons, scholars were able to construct an image of "feudal-bourgeois" conflict that concealed the profound interconnections linking political and commercial elites as uncomfortably entwined elements of a privileged hegemonial "class."

This Marxian analysis seems to have attracted Japanese scholars initially because it enabled them to critique unwelcome aspects of the changes Japan was then experiencing. More basically, perhaps, it offered a professedly universalistic interpretation of Japan's history that enabled them to overcome the debility of being "Asian" in an age when global literate culture in general and historical scholarship in particular were dominated—as so much still is—by Eurocentric Western (white?)-supremacist convictions. It also obviated problems of credibility in the official emperor-centered historiography that the government in Tokyo was promoting as another alternative to Western-supremacist thought.

The Marxian perspective had the serendipitous effect of providing official and radical historians with common ground. The former, as celebrants of Meiji virtue and modern triumph, welcomed the argument that Tokugawa rulers were reactionary, exploitative, and doomed to fail, even if they rejected the Marxian interpretation of why that was so and of what followed. Subsequently, especially in the years after 1945, this Marxian analysis provided an explanation of Japanese militarism, imperialism, and catastrophic defeat that suggested in acceptable ways what had gone wrong and why.[5] By midcentury, therefore, a general consensus existed that the Tokugawa era had been a grim and hateful one.

Other factors made Marxian analysis attractive to foreigners writing in English. During the 1920s–1940s, Europe was convulsed by a fascist-communist rivalry that showed up most harshly in the Spanish Civil War of 1936–1939. Many intellectuals became involved in the ideological struggle, and it even found faint expression in the few notable pre–World War II English-language works on Tokugawa history.[6] Axis defeat in 1945 refocused the fascist-communist rivalry in the form of the cold war, and that change drew most English-speaking intellectuals into the anti-Marxist camp, although a later generation of scholars, distressed by the American military entanglement in southeast Asia, found in Marxian analysis an approach to historical dynamics that made sense in the world they faced.[7]

In English-language studies of Japan, the cold war tensions manifested themselves most influentially in the historicist paradigm known as modernization theory. In its economic aspect, modernization theory prompted closer examination of those who engaged in commerce, and this examination made clear the diversity of the Tokugawa merchant "class" and revealed that its composition, social position, and character changed as time passed.[8] In boldest formulation this view saw Japan's early modern history not as an unhappy story of capitalist forces replacing feudal serfdom with wage slavery but as a favorable story of progress attributable to the dynamics of the "free"—that is, unmanaged, or at least ineffectively managed—market. This view stood the moral ascriptions of the Tokugawa Confucian formulation on their head but modified the Marxian perception mainly in terms of the social value attached to the relabeled bourgeois triumph. In this view the forces of unregulated entrepreneurial enterprise were gradually destroying the constraints of less efficient, authoritarian political and economic arrangements, in the process raising living standards, preparing Japan for industrialization, and creating conditions favorable to the emergence of a "modern" society.[9] In the spirit of its cold war context, this formulation served to validate the moral and historical claims of laissez-faire economics and pluralistic democracy.

By the 1970s, Japan's ongoing economic boom was encouraging Japanese scholars to adopt a more favorable attitude toward their own national legacy, and they found in modernization theory an approach to early modern history that helped explain the socioeconomic dynamism of their own day while transforming militarism and empire from the defining characteristic of modern Japan into an unpleasant aberration. It did so, moreover, in terms that appeared to be as universalistic, and hence as non-Eurocentric, as those of the Marxian analysis they were repudiating.[10]

Modernization theory emerged in a "pre-ecological" day when industrialization, at least in its Anglo-American form, was still viewed as a self-evident blessing—that is to say, no one was asking "what went wrong" that led Japan down the garden path of industrialization. It also was a time when Japan was perceived as unique, the only "non-Western" nation to be "successful" in industrializing.[11] As both of these premises now appear flawed, this body of scholarship is under attack. One line of criticism questions the dichotomous treatment of "government" and "entrepreneurial" enterprise—categories that loom large in the present volume—by noting that early modern merchants and governments collaborated in many ventures. Such collaboration has long been acknowledged for the early Edo period, but scholars are now noting its persistence, indeed its invigoration, in many daimyo domains in later decades.[12]

Modernization theory has been more sharply criticized on the ground that as applied to the Edo period it has disregarded or explained unpersuasively the massive evidence of public hardship, which allegedly belies the notion of rising standards of living or improved conditions of life. These critics have argued, in effect, that later Tokugawa changes benefited some, but at the expense of too many others. To demonstrate and account for these heightened disparities, it is held, the examination of Tokugawa economic history must take into account other aspects of the story, in particular the linkages between and among demography, public health, environmental changes, and popular unrest. This critique has found support in a number of valuable monographic studies.[13] However, no integrated thesis to replace the Marxian or modernizationist formulations has yet appeared.

The absence of a new integrative paradigm reflects the changing context of scholarship. The Marxian and modernizationist constructs were attractive in their day because they explained Tokugawa history in terms that linked it plausibly to concerns of the moment. Today, however, the angst of defeat is a distant memory for Japanese scholars, and even the satisfactions of economic boom are cooling, and among scholars writing in English the cold war no longer spurs serious intellectual activity one way or another. Instead, new issues loom large on the global horizon, and the most fundamental of them pertain to human-environment relations. As these issues force their way into academic consciousness, they shape the way we think about the past as well as the present.[14]

Under these changed circumstances the Tokugawa commercial experience may now seem to make more sense if we adopt a more fully articulated "ecological" approach to economic history, broadening the field of inquiry to encompass more of the variables that determined the costs and benefits of material production and to look more closely at how those costs and benefits came to be distributed as time passed.[15] In particular, rather than assuming that the environmental context was fixed and can, therefore, be ignored, one assumes that it was changing—both shaping and being shaped by human activities—and that those changes were crucial variables in the Tokugawa economic experience.[16]

The story line that emerges from this approach does not appear strikingly new; indeed, the overall rhythm seems akin to that of the Marxian approach. The main differences lie in the explanations of this rhythm and in the broader implications one may find in the story.

The Early Modern Economy: An Overview

One can adumbrate this economic history in the following way. In broadest terms it consisted of two phases. The first was an era of brisk and wide-ranging growth that accompanied and followed military pacification of the realm in the 1570s–1580s. During this era, which lasted through the seventeenth century, the population expanded rapidly, with a postulated twelve million people in 1600 becoming an estimated thirty million by 1720. Gross agricultural output rose commensurately as more and more land came into tillage. Rulers and villagers energetically built and elaborated irrigation systems and leveled and diked fields to grow more wet rice, which they prized for its exceptionally high yield per acre. Cities and towns also grew rapidly, creating a large consumer populace that was provisioned by merchandisers who elaborated the necessary commercial arrangements as decades passed.

As the seventeenth century waned, however, growth slowed, and by 1720 the rise in population had ceased. Little additional arable land could be reclaimed, and little converted from dry grain to paddy culture. Some towns continued to grow, Edo most spectacularly, but most did not, and many lost population, Osaka most notably. In regional terms, the far southwest continued to experience modest growth, the northeast an absolute decline in population. In overall terms the era of growth had given way to an era of stasis, during which producers compensated for an array of intensifying resource scarcities and environmental difficulties by devising ways to improve efficiencies of production. Although their efforts failed to sustain the process of growth, they did avoid severe economic contraction and the human agony and political turmoil that would have come with it. Instead, producers and purveyors struggled with one another to control the products and consumers that were available, and the consuming public, whether migrant laborer or mighty lord, competed and connived to protect their access to the goods and services they desired.

If this overall formulation proves attractive, it will not be because it helps explain Japanese militarism and defeat (a primary concern of Marxian analysis) nor because it helps explain Japan's presumed good fortune in industrializing (the purpose of modernization theory). Rather, it may appeal because it gives this history meaning in terms of the broader global issues of today. It suggests that the island realm of early modern Japan was capable of accommodating a certain amount of social growth, given the material and social technology of the day, but that beyond that point further growth became much more difficult. It required more intensive exploitation of society's resource base, including human labor, and such extension of that resource base as was possible. Moreover, increases in yield commonly were negated by increases in the cost of its production, whether exacted immediately or over the longer term. Even when real gains were achieved, the task of distributing scarce goods in ways minimally acceptable to the populace became more difficult and was accomplished in ways that led to heightened social contention and escalating violence. The story thus does not purport to explain Japan's modern history, but at a broader global level it has about it a contemporary ring that provokes a disconcerting sense of familiarity.

Closer examination of the phases of growth and stability adumbrated here will refine this overall picture of the Tokugawa economic experience. In the process the place in it of commerce in general and forest industries more particularly will be clarified.

Commerce during the Era of Growth

For about a half century after 1580, commercial growth was to a considerable extent driven by political demand. Toyotomi Hideyoshi, Tokugawa Ieyasu, and the other powerful figures who imposed order on the country pursued ambitious agendas of engineering and monument building, in the process putting huge numbers of people to work on projects that created a nationwide capital infrastructure.[17] Commanding merchants, craftsmen, laborers, and villagers, they promoted the construction of irrigation systems, waterways, seaports, fleets of boats, wharves, bridges, roads, highway stations, warehouses, barracks, administrative buildings, mansions, palaces, castles, temples, and shrines. In the process they gave powerful impetus to countrywide primary production, including lumbering.

Besides construction, the pacifiers encouraged commerce by bringing bullion mines under their control, promoting their exploitation, and minting the gold and silver (and later copper) to provide a semistandardized trimetallic coinage for use in commercial transactions.[18] They also standardized measures of weight, length, and volume, mainly to improve efficiency in taxation but with major side benefits to commerce, notably the lumber industry.

The merchants who participated in this government-directed commercial activity functioned essentially as quartermasters *(goyō shōnin)*. Serving as retainers of their lords, they provided mercantile expertise and arranged and oversaw construction and provisioning work in return for stipend and status. With the approval or acquiescence of their lords, they also "piggybacked" entrepreneurial activities on their official functions. They marketed goods such as surplus lumber for a profit or commission and in many instances used their capital to help finance subsequent ventures on behalf of their lords and themselves. Furthermore, the knowledge and connections·they acquired as quartermasters helped them secure large roles in the entrepreneurial sector of the economy as it developed during the first half of the seventeenth century. Most notably, they became major figures in domestic rice and timber marketing, money changing, and foreign trade.[19]

By the 1630s the surge of government-inspired construction was waning, and as it waned the government role in economic life diminished. By then, however, a large population of samurai and commoner families filled some two hundred castle towns scattered across Japan.[20] Being densely settled and mainly built of wood, the towns repeatedly burned, necessitating a high rate of rebuilding and replenishing of household goods. To meet the daily needs of this urban populace, entrepreneurs stepped into the areas of activity that government seemed less and less interested in filling. Consumption diversified, moreover, and as it did, entrepreneurial craftsmen and merchants became the providers of the desired goods.

As entrepreneurial provisioning expanded, merchants found themselves increasingly buffeted by the vagaries of the marketplace. Unprotected by government sponsors, they tried to cope by improving their control of the market, to that end articulating both external and internal aspects of commercial organization.[21] Externally they combined with other firms to form guilds or business associations *(nakama, kumiai)* that could regulate and dominate markets, and the guilds developed rules and regulations as needed to guide the conduct of their members. Within their firms they worked out systems of personnel recruitment, training, and promotion. They also developed codes of conduct for firm members and devised operating procedures to maximize business stability and effectiveness.[22]

The hegemonic Tokugawa regime *(bakufu* or shogunate) in Edo initially opposed such arrangements as *nakama* in the belief that they fostered higher prices and indecent merchant enrichment. Merchants, however, collaborated with one another as best they could, and by midcentury heightened levels of disorder in the marketplace were persuading the rulers to let them form cartels in certain areas of commerce as a way to stabilize the quality, quantity, and price of goods and services. In this gradual shift of policy the rulers first accepted, indeed in some cases created, cartel-like arrangements in areas of commerce most immediately important to members of the samurai caste, such as silk goods and old gold and copper. Only later did they accept them in the major markets, notably grains and wood products, but gradually merchant cartels came to prevail throughout the commercial economy.

In terms of overall structure the commercial system that took shape during the seventeenth century can be characterized as a bipolar central marketplace with feeder links reaching out to the castle towns and from there to the hinterland.[23] The yield of both arable and untilled acreage flowed from hinterland to castle town, partly as tax produce and partly as market goods. Some of this yield was consumed directly in the castle towns, but daimyo and merchants sent a portion onward to the larger cities, notably Osaka and Edo.

Osaka was the primary external destination for produce from western Japan, much coming by boat through the Inland Sea.[24] From Osaka some goods were sent on to Edo to support its million or so people, while some went up the Yodo River to the four hundred thousand in Kyoto, and some sustained Osaka's own comparable population. The tax goods that daimyo marketed in Osaka generated income much of which they transferred to Edo. There it paid for the permanent mansion establishments they maintained in accordance with the shogun's hostage system of *sankin kōtai* (alternate attendance).[25] Daimyo in eastern Japan marketed goods directly in Edo, using the proceeds to cover their costs of alternate attendance.

The transactional loops between hinterland producer and Osaka-Edo marketplace were closed in either of two ways. In the case of tax goods, the cycle was completed, as in our own society, by the government's provision of social order and safety, emergency aid, claims to a benevolent ruling function, and threats, real or feared, to the scofflaw. In the case of entrepreneurial activity, the payments that buyers made to sellers via mercantile middlemen completed the loop.

These middlemen constituted an elaborately articulated subsystem. They specialized by region, type of produce, and area of market activity. Some functioned as purchasing agents, others as transporters, wholesalers, bankers, distributors, or retailers. As the *bakufu* gradually approved cartelization, it generally accepted the patterns of market segmentation that entrepreneurs had developed in preceding decades or even extended them to cover additional areas of trade. As a result the earlier marketing apparatus continued to function into the latter half of the Edo period.

COMMERCE DURING THE ERA OF STASIS

The *bakufu's* ad hoc acceptance of cartelization during the later seventeenth century may have helped stabilize the marketplace; at least the merchant clamor for protection seemed to abate for a while. Economic disorder became more pronounced during the 1690s, however, and as problems of scarcity and public hardship became acute in the 1720s–1730s, a reform-minded shogunal regime gave formal recognition to cartels in a wide array of businesses.[26] It did so by allowing, or even instructing, them to restrict participation in their area of commerce to holders of seats *(kabu)* in their cartel. The move was designed to address urban socioeconomic problems of the moment, but it did so in a way that strengthened the hand of larger-scale urban merchants against any other entrepreneurs, urban or rural, who might wish to enter the marketing arena dominated by cartel members.

For a century thereafter the *bakufu* generally favored the established merchants, probably because they were a known and exploitable quantity. They did, after all, provide the shogunate with many goods and services, including construction, routine maintenance, and repair of such urban facilities as canals, wharves, bridges, and fire towers; funding of community festivals *(matsuri);* and a variety of customary fees, emergency loans, and special grants. In many castle towns, as well, entrenched merchants enjoyed daimyo favor in return for diverse services. Rural entrepreneurs, however, were spared the cost of most such obligations and thus argued that they could provide goods at lower prices. Because the rulers were essentially consumers, this argument proved more and more effective as treasuries grew ever more bare, and during the early nineteenth century such entrepreneurs gradually won government recognition of their right to process and market goods in areas previously monopolized by cartels.

Why rural entrepreneurs could provide goods more cheaply is not entirely clear, but one can point to several considerations in addition to the urban business costs noted above. Because rural merchants were less thoroughly locked into established provisioning routes and procedures, they could minimize middleman costs.

They also were less exposed to fire losses than merchants situated in cities. Their work force was less vulnerable to the illness fostered by the poorer sanitary conditions and greater exposure to communicable disease that characterized city life.[27] Their workers were also less dependent for subsistence on marketed goods, being more able to produce some of their own foodstuffs and other necessities. Added together, these factors placed rural merchants at a competitive advantage, which in the end enabled them to expand their role in provisioning the cities.

More fundamentally the emergence of rural entrepreneurial vigor seems to have been driven by those changes in ecological context that produced the general pattern of later Tokugawa stasis. At the most basic level, the overall shift from growth to stasis reflected a general shift in the balance of supply and demand for an array of primary materials. This shift benefited rural residents because in most instances they were nearer the sources of supply and better situated to locate or develop substitute goods, whether it be fish meal for fertilizer, ground coal for fuel, plantation stock in place of old-growth timber, or domestic silk floss to replace imports that were lost as exportable metal became scarce.[28]

In terms of the food supply, the scarcity of reclaimable land and irrigation water limited cultivators to the arable they had, prodding them to intensify agronomic methods as a means of enhancing output per acre.[29] Increased use of fertilizer was a major element in that strategy, but scarcity of green fertilizer materials resulting from intensive exploitation of woodland and waste raised fertilizer costs enough to spur the adoption of commercial nutrients, notably urban refuse and fish meal.[30] To purchase commercial fertilizer, producers needed cash; but because much of their rice crop went to pay taxes or defray debt, they needed other sources of cash income. So they turned to raising cash crops, further exploiting nearby forest resources, pursuing domestic by-employments as night and off-season work, and seeking off-farm work opportunities.[31] As these practices developed, some cultivators did well, acquiring arable and forest land from others, amassing capital reserves, and becoming landlords, moneylenders, and operators of local commercial ventures. Appreciably larger numbers, however, fared poorly and devolved into tenant farmers and landless laborers, who were more mobile and hence more available for wage work at sites and under conditions selected by employers.

These several trends all contributed to the commercialization of the hinterland, deepening village contact with the wider world. That contact fostered rural literacy and the spread of commercial know-how to the more fortunate villagers.[32] By the later eighteenth century these developments were enabling market-oriented rural producers to undertake more and more of their own processing, transporting, and marketing activity. Exploiting the advantages noted above, they bypassed urban merchants and as a group became influential participants in the country's commercial economy.[33]

Higher up the social hierarchy, the ecological constraints of the later Edo period impinged most visibly in the form of fiscal duress.[34] Tax collection became more difficult, prompting rulers to cut costs, mainly by reducing the stipends of their samurai. Rulers also sought new sources of income. With little more exploitable land available, they had to maximize the tax load on currently assessed output or exploit other forms of production. They did both as best they could, imposing taxes on more and more goods and services, developing more government-controlled commercial ventures, and engaging in more and more forms of monetary manipulation. These efforts commonly involved merchant-government collaboration, but while a few paid off handsomely, many failed miserably. Some created fiscal disorder, some soured merchant-government relations, and some brought rulers into collision with one another. Nearly all such efforts added to the burden of the producer populace, sparking resistance and deepening the hardship of many. However, such efforts, by increasing their tax burden, also gave producers added inducement to improve efficiencies of production and distribution, which tendency favored rural entrepreneurs.

In the outcome the era of stasis witnessed substantial changes in Japan's commercial life. The process of coping with basic scarcities had fostered widespread monetization, promoted countrywide diffusion of commercial know-how and activity, eroded urban control of the market economy, elevated public awareness of social inequalities and vulnerabilities, and complicated the rulers' task of sustaining the Tokugawa political and social order. One can even argue that by the midnineteenth century these developments had primed Japanese society for radical new departures.

Much of the above picture of Tokugawa commerce is reflected in the history of its lumber industry, as the following chapters reveal. There are some variations in particulars, of course, but these too are instructive. The next chapter sketches lumber provisioning in broad terms, indicating the overall character of the process and how practice changed as time passed. Chapter 3 examines the mechanics of timber transport, which was the most costly and difficult aspect of the provisioning process. The fourth chapter focuses on the commercial arrangements that lumber provisioning entailed when handled by entrepreneurs, as exemplified by practice in the Yamaguni vicinity north of Kyoto. The conclusion, or "Last Reflections," then relates this material to the issues adumbrated in the Introduction and to themes sketched here.

Plate 1. **A chute of peeled logs.** Workmen have laid out logs to form a chute *(shura)* for bringing cryptomeria *(sugi)* timber down a V-shaped declivity. After workmen using pole-hooks have dragged and slid out the last logs, they will send on the sticks that form the chute, starting from the upper end. The sticks have been peeled to make the timber slide more easily and to save the bark, if it is marketable, for roofing or other purposes. Reproduced, courtesy of Dean John C. Gordon, from a set of photographs in the Yale School of Forestry and Environmental Studies. These photographs, which were taken during the Meiji period at an unidentified time and location, portray aspects of cryptomeria forestry essentially as practiced during the Edo period, except for the use of crosscut saws.

Plate 2. **Splash dam and chute.** This splash dam holds back enough water to
float logs to the dam's face. There men equipped with pole-hooks pull them
one at a time onto the chute *(shura)* that guides them through the ledge
outcropping to the next stretch of pooled water, where logs have already
drifted to the face of the next dam, as well as the one beyond it. The log-
and-pole frame resting on the ledges below the chute gives workmen a
place to stand when dislodging logs that jam among the rocks. Reproduced,
courtesy of Dean John C. Gordon, from a set of photographs in the Yale
School of Forestry and Environmental Studies.

Plate 4. **A streamside assembly point.** At the foot of a trestle *(sade)*, upper
right, peeled logs pile up at an assembly point on a river large enough to
handle rafts. Workmen use their pole-hooks to drag the logs onto the temporary processing dock, drill the holes through which binding vine will
pass, and then roll the sticks into the water, where they are assembled into
raft sections. These Meiji-era logs have been cut by crosscut saw. Reproduced, courtesy of Dean John C. Gordon, from a set of photographs in the
Yale School of Forestry and Environmental Studies.

Plate 3. **Wood-processing site.** Felled and limbed trees at this site are cut to proper lengths, split into shingling or sawn into board stock, and stacked for curing before shipment by pack animal, at least as far as a raft landing where the pieces will be forwarded by water as raft cargo. The crudely built workmen's bunkhouse is protected from runoff flooding by the rock-faced levee in foreground. Reproduced, courtesy of Dean John C. Gordon, from a set of photographs in the Yale School of Forestry and Environmental Studies.

Plate 5. **A full-sized raft and crew.** A raft crew of seven men guides a raft of about twenty sections along a quiet strip of water below precipitous hillsides. Such a large crew suggests that other parts of the journey are less tranquil. The high waterline on the far riverbank and the accumulation of large stones on the near bank attest to the size and power of the river in flood. Reproduced, courtesy of Dean John C. Gordon, from a set of photographs in the Yale School of Forestry and Environmental Studies.

Plate 6. **A cryptomeria plantation.** At right a well-thinned and trimmed stand of maturing cryptomeria with a rich understory of low growth faces a sunlit patch of hillside recently clear-cut and replanted to a new crop of trees. Reproduced, courtesy of Dean John C. Gordon, from a set of photographs in the Yale School of Forestry and Environmental Studies.

Plate 7. **A hillside logging site.** A cryptomeria stand on a hillside above a ledgey streambed is gradually clear-cut. Loggers fell the trunks uphill, top and peel them, and cut them to lengths. They then split the logs into shingling, which they stack to dry on the temporary racks they have erected before they carry the pieces down to the river. The footpath down the hillside zigzags from one stack to another and emerges from the woods at bottom center, where bridging leads down to the riverbed. Reproduced, courtesy of Dean John C. Gordon, from a set of photographs in the Yale School of Forestry and Environmental Studies.

2 | AN OVERVIEW OF LUMBER PROVISIONING

LUMBER PROVISIONING was a major industry because early modern Japan's several hundred towns and cities were wooden creatures with ravenous appetites for timber. The character of those appetites changed as time passed, moving through two distinct phases. The first, from the 1570s up to about 1650, was a "boom" phase in which the realm engaged in a vast amount of building. It involved a superabundance of monument construction—mainly castles, palaces, mansions, temples, and shrines commissioned by the newly consolidated samurai ruling elite—and a surge of urban construction that dotted the realm with population centers. These ranged in size from the great cities of Edo, Kyoto, and Osaka down to a large number of towns containing a few thousand residents each.

The decades of boom were followed by a long "maintenance" phase from about 1650 to the midnineteenth century. During this phase society tried by ceaseless repair and replacement to maintain the existing urban plant. Except in Edo, whose population kept growing for another century, the effort added little that was new and probably failed even to sustain either the quality or scale of initial construction.[1]

During both of these phases three methods of wood provisioning were practiced, each a facet of one of the three economic systems—subsistence, command (government-directed), and entrepreneurial—that characterized the broader economy.[2] Most of the following discussion focuses on command and entrepreneurial lumbering because they provisioned the cities, but because the three methods overlapped in practice, subsistence lumbering merits a brief note.

Subsistence lumbering, in which the producer was also the processor and consumer, is the least well documented of the three. In it, villagers, who constituted about 80 percent of the population, went into nearby woods, felled trees, processed the wood, and carried it home for use. Such production was important in itself, and it also functioned as a marginal element in the other lumbering systems. Most notably, government and entrepreneurial loggers sometimes defrayed labor costs by allowing locally hired workers to harvest designated areas of woodland for home use or take the scrap wood left over from organized logging projects.

Whatever the particulars, subsistence lumbering met the shelter requirements of many people and probably moved the largest volume of wood.[3] It was capable only of meeting local needs, however, and it achieved its overall significance solely through vast multiplication of simple, localized operations. Command and entrepreneurial lumbering, systems organized at the regional and national levels, provided the materials for monument builders and city dwellers.

Command Lumbering

Command or government-directed lumber provisioning, that of both the *bakufu* and the 250-odd daimyo domains *(han)*, is the most richly documented type of lumbering.[4] It was most important during the boom phase, being central to both monument construction and the initial building of most towns, and in one form or another it persisted as a source of government income and replacement timber until the end of the Edo period.

There were two basic types of command lumbering: sporadic or special-order provisioning for use or sale and regularized production that was usually intended to generate government income. In the first type a ruler ordered a particular project undertaken and instructed officials or vassals to obtain the necessary material, the wood being known generically as *goyōki* (lord's wood). A few examples will illustrate the type.

SPECIAL-ORDER PROVISIONING

During his years as hegemon Toyotomi Hideyoshi obtained timber for his great construction projects via special orders. In response to one such order in 1593, Kyōgoku Takatomo, the daimyo of Iida han on the Tenryū River, furnished lumber for the construction of Osaka castle, sending the material downstream to Kakezuka and thence by ship to the building site.[5]

Similarly Akita Sanesue, lord of Kubota han in the far north, provided Hideyoshi with lumber for several projects. Sanesue's workmen rafted the required wood down the Omono and Yoneshiro Rivers to ports where it was loaded on ships, hauled to Tsuruga, portaged to Lake Biwa, and rafted to the construction site, a trip of nearly two months' duration. In 1593 Sanesue provided one boatload of ship's planking as ordered and the next year thirty boatloads to fabricate vessels for use on the Yodo River. In 1595, when Hideyoshi was constructing Fushimi castle, Sanesue shipped timber of unspecified dimensions that totaled some six thousand feet in length. Hideyoshi ordered him to provide a specified volume of four-inch plank in 1596; and in the three following years, he submitted annual orders for a fixed volume of shorter five-inch plank.[6] Thus before Hideyoshi's death in 1598, he seemed to be moving from a system of special orders to one of regular requisitioning.

In lumber provisioning, as in so many things, Tokugawa Ieyasu's practices were similar to Hideyoshi's. In 1606 he ordered the several daimyo in Shinano to send shingling materials to Edo for use in castle construction. Each daimyo was instructed to furnish specified quantities, the volume proportional to the putative agricultural yield *(kokudaka)* of his domain. The lords hired woodsmen to fell the trees and reduce them to portable size and stablers to transport them by packhorse overland to Takasaki. From there raftsmen took them down the Tone River to Edo for storage at the government lumberyard just outside Ōte gate on the castle's east side. This order yielded some forty-three thousand pieces of wood produced by 144,550 man-days of labor.[7]

The same year Ieyasu's intendant in charge of the Kiso River valley provided an additional 47,668 pieces of shingling wood for Edo castle. He also furnished specified numbers of precisely defined items (including 943 pillars eighteen feet by twenty-one inches square and 5,996 pillars eighteen feet by six or seven inches square, all of *hinoki*) for Sunpu castle, which the newly retired shogun was enlarging.[8] The intendant had the pieces prepared and floated down to Ise Bay, from whence they were shipped to work sites in Edo and Sunpu.

In later years, as government construction tapered off, lords used their forests as a source of emergency income. In 1621 the daimyo of Tosa resolved to liquidate his government debt by selling timber. He appointed two officials to oversee the project and ordered one hundred minor vassals to the work site to carry it out, later assigning another hundred to the task. They and their workmen felled the timber, cut it into five- or six-foot lengths, and floated it eastward down the Yoshino River to the coast. From

Figure 1. **Kiso loggers relaxing in their bunkhouse.** Each man has his own sleeping mat in front of the bank of fires that heat water, cook food, and warm the hut. Reproduced from Tokoro Mitsuo, *The Wood-Cutting and Transporting System of Kiso* (Tokyo: Tokugawa Institute for the History of Forestry, 1977), scroll 1, plate 3. Courtesy of the Tokugawa Institute and its director, Ōishi Shinzaburō. This painting was originally produced during the 1850s as part of a two-scroll set by Tomita Ayahiko (1811-1877), a petty functionary in the *bakufu* intendancy at Takayama in Hida Province. The forty-two paintings in the set illustrate the process of Kiso lumbering from initial felling to transport of the wood to Edo.

there it was shipped across the Inland Sea to Osaka and stored at a waterfront site assigned to Tosa by the *bakufu*. The *han* then sold the wood to city lumber merchants for a sum that erased its debt and left money in its vaults.[9]

In their special-order provisioning, rulers routinely used merchants, sometimes to organize the logging, usually to arrange the transporting and storing, and often to sell any timber produced for government income. Initially the merchants served their lords as *goyō shōnin,* with their own entrepreneurial activities ancillary to the acquisition of government timber. Later, however, governments relied more fully on merchant entrepreneurship to meet sporadic needs.

One example: In 1680 the *bakufu* ordered Numata han in the northern Kantō to furnish timber for building Ryōgoku Bridge on the Sumida River.[10] Numata officials arranged to have the Edo lumber merchant Yamatoya handle the task, granting him authority to fell trees in *han* woodlands along the Tone and Katashina Rivers. Evidently Yamatoya was allowed to cut out the equivalent of eight thousand large timbers, the best three thousand going to the *bakufu* as *goyōki* for use in building the bridge. In return for the remaining wood, which he could sell at market, Yamatoya was to pay Numata two hundred *ryō* in commission fees. The contract specified the areas to be cut, limitations on choice of trees, procedures for payment, and completion date. Yamatoya hired woodsmen to convert the trees to timber and local villagers to move the pieces to the river and downstream to a landing where water level permitted raftsmen to assemble them for the float to his lumberyard in Edo.

REGULARIZED PROVISIONING

In addition to sporadic provisioning of the sort exemplified above, rulers regularly exploited their woodlands to meet routine timber needs and, more commonly, to generate income. *Bakufu* and *han* leaders often allowed—or required—residents of wood-producing areas to pay their taxes in timber or other wood products instead of the usual rice. The taxpayers got out the wood and arranged to float or otherwise transport it to an official receiving area. There it was measured, recorded, and stored; forwarded to another government-operated depot; or turned over to a merchant for sale, the proceeds going to the treasury.[11]

The rulers also obtained *goyōki* by having their bureaucracies

Figure 2. **Ryōgoku Bridge in Edo.** The view is from upstream on the east bank of the Sumida River looking southwest across Edo toward Mt. Fuji. The timber, labor, and engineering requirements of this major bridge were substantial. The inflowing Kanda River is seen beneath the small bridge to the right of Ryōgoku. The foreground junk is transporting a potpourri of monks, merchants, and peddlers across the river, while the woman at lower left does laundry. Reproduced from an unidentified partial edition of the series of prints "Thirty-six Views of Fuji," by Hokusai Katsushika (1760–1849), in the author's possession.

manage the logging directly. Initially they linked such lumbering to the tax system by impressing local people as corvée laborers and forgiving them other tax assessments, but by the eighteenth century they commonly paid workmen daily wages. This type of lumber provisioning was widespread, two well-studied cases being Owari han's logging on the Kiso, which is treated in chapter 3, and Kubota's logging of the Yoneshiro watershed.[12] Here Hiroshima and Hitoyoshi han will exemplify the practice.

Hiroshima's arrangements constitute a simple version of government-operated lumbering.[13] The wood moved via the Ōta River, which traverses Hiroshima city en route from interior highlands to the sea. Logging sites along the river were comparatively easy of access, so professional loggers were unnecessary; ordinary villagers handled all tasks from felling to delivery. Near year's end, *han* finance officials prepared an estimate of total government timber (and fuel) needs two winters hence. Villagers commenced the logging after receiving estimates of the quantities they were to provide and an advance on their wages.

Late in the winter the finance office reassessed its wood needs and had forestry officials send final requisitions to intendants. The intendants forwarded them to local forest inspectors, and thence to village heads, who were to assure that each village met its obligation by autumn. Villagers continued working at the task as time permitted. When autumn came, local forest inspectors or subordinate officials in the forestry office traveled to the villages, examined the yield, stamped pieces with the *han* seal, and ordered village heads to store the wood securely over winter while it cured. A few days before year's end officials made a final tally and paid villagers the rest of the contractual sum (provided a village had met its obligation), also issuing at that time their estimates of the subsequent year's lumber needs. After winter passed, villagers organized the *goyōki* into rafts, which they floated downstream to Hiroshima for disassembly, storage, and eventual use or sale.

Another well-documented case of regular command lumbering is that of Hitoyoshi han in Kyushu. Hitoyoshi's topography was more forbidding than Hiroshima's, and logging there involved more complex arrangements. Officials in the timber office made an annual estimate of government wood requirements and drew up a general logging plan.[14] After senior officials approved the proposal, it was sent to the forestry office, where foresters selected cutting sites and dispatched assistants to work with village woodland overseers in marking trees for felling.

The marked trees came under the jurisdiction of an official from the timber office, who became personally responsible for supervising the crews of expert loggers who were hired individually and the gangs of day laborers provided by nearby villages. The former did the actual felling and processing; the latter employed chutes, sleds, and draft animals to work the wood downhill to an assembly place near the river. From there other unskilled workmen, usually corvée laborers from villages abutting the river, took over the transport work. Their tasks were to move the wood downstream via free float to a landing at Hitoyoshi town and to assure that none was lost or stolen en route. At

Hitoyoshi the pieces were collected and bound into rafts, which professional raftsmen steered down the Kuma River to Yatsushiro Bay for loading onto ships bound for Nagasaki, Osaka, or other destinations.

Evidently problems arose in these arrangements because Hitoyoshi eventually modified them. In 1775 the *han* began paying its corvée laborers because river work fell inequitably on the local populace and generated discontent. Eleven years later, in an apparent attempt to trim its increased transport costs, it shifted to contracts that specified piecework rates for moving given volumes of wood from the assembly place to Hitoyoshi and invited villages to bid for the work. The *han* also began selling standing trees to Hitoyoshi merchants, thereby avoiding the whole managerial problem. Subsequently, in 1845, it began contracting with merchants to handle on commission all *han* lumber sales in Nagasaki, Osaka, and other market centers.

As these several cases indicate, in command lumbering, initiative came from political authorities. Officials decided whether and when to log, and political bodies were more or less centrally involved in implementing the projects. Work procedures varied widely, however. As the Numata and Hitoyoshi instances suggest, government and entrepreneurial logging could become thoroughly entangled, with merchants assisting rulers—and themselves as best they could—by both providing and disposing of government lumber.

Entrepreneurial Lumbering: Some Examples

The merchants who worked for rulers also engaged in felling and selling activity that had little or no relationship to the latter's timber needs. That work, entrepreneurial lumbering, constituted the third major pattern of timber provisioning.

Popular wisdom, which says that Tokugawa-period lumbermen became obscenely wealthy, seems credible in the two famous cases of Naraya Mochizaemon and Kinokuniya Bunzaemon. Naraya, who flourished during the 1630s, provided lumber to construct the Tokugawa mausoleum complex (Tōshōgū) at Nikkō and reportedly left his heir four hundred thousand *ryō* in treasure.[15] A few decades later Kinokuniya (1669–1734), who was born in Kii Province, reputedly made his initial fortune shipping mandarin oranges eastward to Edo and fresh fish westward to Osaka and Kyoto. Subsequently he moved to Edo and set up a lumberyard. When a conflagration erupted one night and spread through the city, he reportedly sailed west even before the fire had died down to purchase huge quantities of Kiso lumber. When the wood reached Edo, he sold it at vast profit, using his wealth to live a wild life in the Yoshiwara entertainment district. Kinokuniya clearly was a major provider of lumber to the Tokugawa in the years around 1700, but his reputation for high living has surely grown in the retelling.[16]

Doubtless other timber merchants amassed handsome fortunes, especially during the boom years of monument construction when rulers spent lavishly to celebrate their achievements and pretensions. Even in the nineteenth century it was possible to do well in lumbering. Two documented examples, one a small-scale entrepreneur from the Yamaguni area north of Kyoto and one a large-scale merchant in Edo, provide more solid evidence of this phenomenon than do the near-mythical careers of Naraya and Kinokuniya.

NOGAMI RIHYŌE

The Nogami family of Ōno village in Yamaguni was a modest farm household in 1783, with agricultural land assessed at 1.13 *koku* of putative yield.[17] In 1806, when the family's arable land had expanded to 2.14 *koku*, the Nogami acquired a *yama yaku*, meaning they gained control of a plot of woodland with the right to log it and the obligation to pay a tax on the yield. In following decades the family accumulated more land, and when energetic young Rihyōe assumed the family headship in 1831, he was registered as holder of 6.38 *koku* of arable land. That year the production of Nogami woodland, which had yielded a tax of 3.1 silver *monme* in 1806, yielded 4.4. During his years as family head (1831–72), Rihyōe continued acquiring resources, and by the 1860s his arable land had doubled to 13.2 *koku*. His forest output, which had risen steadily to yield a tax of some 8 *monme* by the late 1850s, then shot upward during the 1860s to a minimum of 25 *monme*, in some years producing as much as 60.

From the outset Rihyōe promoted forest work. In 1837 he purchased a seat in the local raftsmen's guild, which allowed him to ship timber down the Ōi River for sale in Kyoto. By the 1840s he had sufficient venture capital to buy standing timber from other woodland holders and arrange its logging and shipping. In this way he continued to expand, and by the early 1860s he was sending about forty raft loads (roughly 30,000–40,000 cubic feet) of timber downstream every year.

During the 1860s Kyoto was racked with political disorder that produced city fires and a high but irregular demand for lumber. Amply funded and well positioned to exploit the opportunity, Rihyōe flourished. As the scope of his activity outgrew his supervisory capacity, he adopted the simpler managerial policy of a timber broker. He provided neighbors with advances on their timber production, let them manage their own logging, and then disposed of the yield for a commission when they delivered it to him. To protect his capital, he accepted arable land or woodland, standing timber, or most often the rafts themselves as security on his advances. Finally, in about 1868, as Tokugawa rule was ending, Rihyōe moved his household from Ōno down to the raft landing at Saga on the western edge of Kyoto. There he could oversee his business better and prepare to move into the marketing end of the industry.

NOGUCHI SHŌZABURŌ

The history of the Edo lumbering firm of Shinanoya is much grander than that of Nogami Rihyōe.[18] It stands midway between Rihyōe's experience, which probably is a fair sample of the typical success story in early modern lumbering, and the thoroughly atypical achievements of the legendary Naraya and Kinokuniya. Shinanoya was the firm name of a certain Noguchi family that resided in a small village in Shinano Province from the 1580s onward. Generation after generation, the Noguchis tilled their soil while producing small pieces of salable lumber from the family woodlot.

Shōzaburō, born to the family in 1799, learned the wood trade as a youth and began expanding his timber activity after taking over the household in 1828. Evidently using family savings as venture capital, he traveled extensively, purchasing timber from landholders in the mountains of nearby provinces and reselling it to lumber merchants in Edo. Frustrated with the city's oligopolistic wholesaler-broker (ton'ya-nakagai) system, which forced him to market through intermediaries, he purchased a lumber broker's license in 1836. The license enabled him to open a city office in the Kanda district, where he could bargain with the wholesalers from both sides, as producer selling to them and as broker buying from them. Thus strengthened, he consolidated his role as a key purveyor of timber from central Japan.

Then in 1844 the labyrinthine headquarters building *(hon-maru)* of Edo castle burned, and Shōzaburō moved briskly to provide reconstruction material.[19] Knowing the woodland in much of central Japan and having contacts there already, he quickly located the 150 timbers, each roughly twenty inches square, needed for basic framing. In a risky and unexpected move that saved weeks in transport time, he had the huge timbers hauled directly over the mountains from landlocked Matsumoto and down the Tone River to Edo. Impressed by this feat, the *bakufu*'s finance magistrate designated him an official lumber merchant. The appointment enabled him to bypass the wholesalers and directly provide large quantities of board stock and other material for the castle work. At government behest his buyers scoured almost the entire region between Edo and Kyoto and traveled to Shikoku and northeast Japan in search of satisfactory timber. The following year, after another city fire, he again brought large quantities of timber to the city, this time for public sale.

Shōzaburō's official appointment expired when the castle reconstruction was completed in 1847, but shortly afterward he acquired a wholesalers' license and established a lumberyard at Fukagawa Kiba. During the 1850s he provided lumber for numerous projects, including reconstruction following two more fires in Edo castle and one at the imperial palace in Kyoto. Besides provisioning the *bakufu,* he took salaried positions as timber supplier to several daimyo. He promoted reforestation to assure future timber supplies and organized some land reclamation projects. In 1857 the *bakufu* rewarded him with "quasi-samurai" status, which allowed him to wear a sword and use his family name in public.

During the 1860s Shōzaburō continued to purchase timber throughout Japan, even as far away as southern Kyushu. He opened branch lumberyards in Nagoya, Osaka, and Kyoto, eventually employing some three thousand people. From 1865 to 1867 he served again as official lumber purveyor to the *bakufu,* but ever alert politically, he responded to the collapse of the

Tokugawa regime in 1868 by transferring his main residence from Edo to Kyoto, headquarters of the civil war victors. There he demonstrated his devotion to their cause by undertaking repair of the imperial palace. Lest affairs develop unexpectedly, however, he also opened a branch office in Sunpu, to which town the defeated Tokugawa had been dispatched. As things worked out, this last prudent gesture proved unnecessary, and when Shōzaburō died in 1871, he left behind a flourishing lumber merchant's household. Like that of Nogami, it was essentially the creation of one lifetime of shrewd entrepreneurial activity.

Entrepreneurial Lumbering: Problems and Solutions

Doubtless the personal drive and business skill that underlay the successes of Nogami of Ōno and Noguchi of the Shinanoya were present in the builders of the more famous Naraya and Kinokuniya firms. Clearly, however, timing also was important: these merchants were able to provide goods just when demand was highest. Moreover, Shinanoya's ties to government, like those of Naraya, underline the attraction for lumbermen of semiregulated government provisioning. Ever since the command lumbering of the early seventeenth century, lumbermen struggled to replicate in the commercial marketplace the order and economic security of government provisioning. Their inability to do so was a source of chronic concern.

Lumbermen yearned for an orderly market environment because, glorious success stories notwithstanding, entrepreneurial lumbering was a precarious business. Shimada Kinzō has noted that despite the ideal of business permanence, timber merchant families in Edo city rose and fell rapidly, almost none surviving more than the "three generations" allotted merchant households by customary wisdom.[20] Indeed, their work was so uncertain that the great majority of them rented rather than owned their city facilities.[21]

It appears that throughout Japan lumbermen experienced this uncertainty, most operating for only a few years and then disappearing from view, temporarily or permanently. Moreover, lumbermen usually maintained concurrent businesses, agriculture and money lending most commonly, but also other forms of forest work, such as producing charcoal and fuel wood. These enterprises generated a more regular income, which helped capitalize logging ventures, absorbed fluctuations in the timber marketplace, and provided a fallback position when business soured.

Capitalizing the Business

Entrepreneurial lumbermen needed ample capital and adequate fallback positions for two reasons. First, lumbering involved a substantial investment up front. Even in the traditional exploitation lumbering of the seventeenth century, in which reforesting was not practiced, the costs of buying the standing timber, paying the logging crews, hiring the shippers and warehousemen, storing the timber, and paying the workmen who hauled the goods to the ultimate buyer were all incurred, in part or in full, before the consumer paid for the goods. Moreover, the consumer was fully as likely to pay in installments as in immediate cash, further extending recovery time on the entrepreneurial investment. Then, during the eighteenth century, as timber depletion gave rise to plantation forestry, lumbermen had to absorb a portion of the long-term investment costs required by the new silviculture techniques.[22]

The second economic risk in the industry derived from the striking irregularity of both supply and demand. Supply fluctuated with the growth cycles of woodland. Storms and forest fires compounded the problem by abruptly producing large quantities of salvage timber that required disposal before it rotted and then by leaving a wasteland that might yield no further timber for decades. Vagaries of weather, such as flooding or severe snow, could also stop shipments, creating temporary scarcities and price hikes. Demand fluctuated just as dramatically because urban conflagrations produced sudden bursts of consumption that lasted until the city was rebuilt, whereupon lumber usage plummeted, gradually reviving a decade or so later as normal replacement needs reappeared. Sometimes, moreover, fires led to price explosions that lured excessive quantities of wood to the city, creating an oversupply that severely depressed the market. Hence lasting success in the business required capital sufficient to accommodate both normal investment costs and sharp swings in supply and demand, and lumbermen found those requirements difficult to meet.

To implement a lumbering project, an entrepreneur had to mobilize and deploy enough capital that participants at all stages of the operation would perform their portion of the overall task.[23] The venture capital could derive from the landholder, local lumberman, shipper, wholesaler, or broker, depending on how badly one or another wished to buy or sell. Had early modern Japan been a capital-rich society, any one of these participants might have underwritten the work, but it was capital poor, so lumbering required the combining of resources. Most commonly, it appears, the several participants all contributed to the venture, either by providing labor in advance (e.g., plantation operators, fellers, haulers, raftsmen) or by capitalizing segments of the operation in advance (e.g., the landholder, logger, shipper, wholesaler, or broker).

Lumbermen developed financing techniques as elaborate as the projects being undertaken. The final marketing usually was a straightforward transaction accomplished through fixed pricing, sealed bids at auction, or direct negotiation between buyer and seller or agent. However, so many parties were involved in the forest-to-city operation as a whole that funding of its several stages habitually entailed elaborate patterns of prepayment, installment payment, commissions, paybacks, and add-on charges. These intricate arrangements enabled participants to spread their costs and risks and helped them bind one another into commitments that maximized the likelihood of the whole sequence of transactions being carried to completion and thus of participants recouping their investment in labor or capital.[24]

Procedures varied from place to place and changed as time passed, but in a stereotypical project, a lumber wholesaler (zaimoku ton'ya) sent his representative into the countryside to negotiate a contract with a timber seller, usually a local land-holding lumberman or lumbermen's guild. The wholesaler offered to dispose of the wood for the best price, but for no less than an agreed minimum, in return for a specified commission plus expenses. The representative might make an advance payment to get the work started or simply to reserve a stand of trees for later felling, with the balance following upon sale of the goods.

With the contract arranged, the seller assembled his work force of professional choppers and skidders, any skilled woodworkers required to convert logs into posts or flat pieces prior to shipping, and the unskilled laborers who handled much of the transportation work. This crew might consist mostly of migrant

Figure 3. **Sawyers at work.** Sawyers at a logging site near Yamanaka in Tōtōmi Province rip a timber to form planks for shipment to city markets. Most timber from the interior of Tōtōmi went down the Tenryū River to Kakezuka and thence by boat to market. The man at lower left sharpens his saw while family members look on. In the background, lower center, a workman rests and enjoys the view of Mt. Fuji to the northeast. Hokusai took artistic license in placing one sawyer directly under the sawdust fall of another. Reproduced from an unidentified partial edition of the series of prints "Thirty-six Views of Fuji," by Hokusai Katsushika (1760–1849), in the author's possession.

professional loggers and unskilled laborers *(dekasegi)* for whom the jobs provided winter work, or, especially in the later Edo period, it might be composed solely of local people.

Fellers and woodworkers did the logging and processing, normally receiving their wages in installments before, during, and at conclusion of the job. Raftsmen and shippers moved their goods as specified by contract, and when the wood reached the wholesaler's yard, he publicly announced what he had for sale. Interested lumber brokers *(zaimoku nakagai)* submitted bids, and the successful bidder had two or three months in which to pay for his purchase. Once the broker had negotiated the resale to contractors, retailers, or government agents, he arranged delivery of the goods from wholesaler's lumberyard to buyer's site. Ideally the buyer paid the broker in time for him to pay the wholesaler. Then the wholesaler, in his semiannual settling of accounts, deducted his commission and fees, recovered any advance plus accrued interest, and forwarded the contractual remainder to the shipper, who took out his fee and delivered the rest to the seller. The seller paid raft operators, settled any debts owed his logging crew, and then discovered how much remained to cover his investment costs and constitute his profit.

A lumber shipment from the Kii peninsula illustrates aspects of the payment pattern.[25] Members of the Kumano lumber wholesalers' guild in Edo handled timber that originated in the precipitous valleys facing the coast near Kinomoto. A local organization of lumbermen, the Ise kō nakama, managed the lumber production, using their own capital to purchase standing trees and hire fellers. Before rafting season ended in the spring, members of the local raftsmen's guild moved the wood from riverside to Kinomoto, where the pieces were stamped with the Ise kō nakama seal, turned over to a shipper, and loaded aboard ship. Accompanying the wood was an invoice that identified the producer, quantity of wood, carrying vessel, and agreed price of transport to Edo (which commonly was about half the total cost of lumber sent to the city from Kii).

Sailing from Kinomoto, the shipper received official clearance at Shimoda and delivered his cargo to the appropriate Kumano guild member's lumberyard in Edo, whence it was sold via a broker. The wholesaler received payment from the broker, deducted his commission and expenses, and in the autumn settling of accounts turned the balance over to a runner, who carried it overland to Matsusaka on the coast south of Nagoya. From Matsusaka a second runner carried the money to Kinomoto, so that it reached the shipper there about twenty-five days after leaving Edo. How and where the runners received their pay is unclear, but before sending the balance upriver to the local lumbermen, the shipper took out his fees and designated expenses as specified by contract, and they may have included the runners' charges. At the village lumbermen's office, raftsmen received payment, and the lumbermen's organization then distributed what remained among its members in accordance with their own prior arrangements.

Thus, even when payment was handled expeditiously, several

months might elapse before an entrepreneur recovered his capital. Often, moreover, the final buyer was unable to pay for his goods immediately and arranged installment payments instead. Or the broker was unable to resell the goods or sold them at a loss and could not meet his obligation to the wholesaler. Under such circumstances recouping might take years.

ESTIMATING THE YIELD

The need for investment capital, irregularity of supply and demand, and problems surrounding repayment posed obstacles enough to commercial lumbering. They were exacerbated by the difficulty loggers faced in assessing the value of stumpage. During the Edo period woodsmen estimated the timber in a forest in various ways. In smaller areas with identified boundaries, they could simply count the trees available for taking and estimate their worth individually. In larger or poorly bounded areas, they "eyeballed" the stand, relying on their knowledge and experience to form an estimate of the number, size, species, and condition of harvestable trees and hence the volume and market value of its lumber.[26]

Under the best of conditions forest mensuration was (and still is) a tricky art, and as timber stands deteriorated and exploitation became more thorough, errors in estimating became both easier and more damaging. One problem was that restrictions on felling multiplied. More and more species came under official control, and authorities also tightened control by size, prohibiting the cutting of trees whose circumference at eye level was above (or below in some cases) a specified figure. Moreover, most logging activity was selective, and earlier harvests commonly left defective trees, whose presence could give a woodland a false appearance of lushness. As stands became thinner, defective trees more common, and cutting more restricted, errors in mensuration became easier because the mensurator had to discount so much of what was before him. Should he overestimate the content of thin and scattered stands, the task of fulfilling his contractual obligation might prove impossible, or at least saddle him with a backbreaking labor cost as workmen hauled small stems out of distant nooks and crannies.

One example will illustrate this peril of entrepreneurial lumbering, even when timber appraisals had official sanction and were made under close government oversight.[27] Villagers along the Koshibu River in southern Shinano were accustomed to supplementing agricultural income by logging nearby *bakufu* woodland. Their work was capitalized and managed by merchants from Funagira and Kakezuka, downstream on the Tenryū. In 1730, during a sharp depression in agricultural prices, the villagers petitioned to log, but not until 1737 did the *bakufu* intendant consider the stand adequate, accept their petition, and obtain instructions on its implementation from the finance office in Edo. These instructions approved the logging plan submitted by Yamagataya Kitauemon, a merchant from Kakezuka. They authorized him, over five years, to fell some thirty-three thousand trees, which were expected to yield fifty thousand *shakujime* (589,265 cubic feet) of timber. They forbade him to fell *sugi* or *hinoki* of any size and trees of any species that would not yield a butt log at least three inches square by six feet long. Kitauemon was to pay a 10 percent produce tax, defined as equivalent to 550 gold *ryō* annually, and an additional forfeit fee should he fail to produce the stipulated tax. His family land at Kakezuka was to stand as surety.

The *bakufu* instructions charged Kitauemon with hiring workmen, processing the wood, and floating it down the Koshibu to Kazurajima landing on the Tenryū. There officials would measure the yield and stamp the best 10 percent—the produce tax—as *goyōki*. Workmen would then bind the wood into rafts and send it on downstream. The *goyōki* was to reach one of the two downstream *bakufu* lumber storage sites within fifteen days of the stamping to minimize delays that could promote rot. Raftsmen would float the remaining timber down to Kakezuka, where shippers would load it and haul it by sea to wholesale lumberyards at Fukagawa, from whence it was to be sold.

Before work started, the intendant dispatched an examiner, two assistants, and an orderly to oversee the project at the felling site. He stationed other officials at Kazurajima to measure the incoming timber and mark the *goyōki*. Timber tax collectors at the two downstream landings received the government wood for storage, noted any damage or delinquency, and ensured that all rafts of merchant wood proceeded to Kakezuka in good order.

The project commenced on schedule, but it soon became apparent that the forest was too understocked to yield the required material. Loggers began cutting smaller trees and shipping out smaller pieces, but even though they worked the seventy-five thousand man-days called for in the plan, they were still unable to produce their first year's 10,000 *shakujime*. Indeed, they extracted only 4,808. Subsequently output increased, but at the end of four years the work still yielded only 27,690 *shakujime*.

Well before then Kitauemon was in arrears on both taxes and wages, and because of the error in mensuration, production costs were outstripping the market value of what he was extracting. In

Figure 4. **Measuring the yield of government wood.** The six workmen standing at center front use pole-hooks to slide and role the squared timbers into position while a sworded official measures each stick. Behind them, at foreground right and center left, two workmen carve identifying marks into the pieces while another at center rear blackens the marks to make them visible. A properly sworded government clerk seated at upper left and a seller's representative squatting at lower left maintain tallies of the measured timber. At upper right the seated man enjoying a smoking break holds the bamboo pole he uses to measure stick length. To his right a happy axeman, having finished squaring timbers with his broadaxe, cools down, while the tea server behind him heats water preparatory to serving the crew. Reproduced from Tokoro Mitsuo, *The Wood-Cutting and Transporting System of Kiso* (Tokyo: Tokugawa Institute for the History of Forestry, 1977), 101. Courtesy of the Tokugawa Institute and its director, Ōishi Shinzaburō.

1739, as he saw his plight deepening, he tried to borrow two thousand *ryō* from a lumber merchant in Kiso, but was refused. In the spring of 1740, although 10,000 *shakujime* were lying about the woods, he halted the work, apparently because he was bankrupt. The villagers evidently went ahead on their own and moved most of the felled timber down to Kazurajima, where they had the intendant impound it as surety on about two years' worth of wages and some back taxes.

With affairs deadlocked, the villagers petitioned the *bakufu* for a revised contract, proposing a 30 percent reduction in total output, a proportional reduction in tax, and a delay in the project's final deadline. Some Edo lumbermen supported the petition, and in 1742 the finance office approved the compromise. Kitauemon returned to work, evidently without incurring the punitive fees of his first contract, and the impounded wood was processed, the remainder taken from the forest, and the scaled-down project finally completed in mid-1743. In the end the villagers earned in seven years of part-time logging work about 70 percent of what they initially had expected to make in five; the *bakufu* received commensurately reduced taxes for a greater investment of supervisory time; Kitauemon escaped with his household intact; and the forest was left more depleted than ever.

Entrepreneurial Lumbering: The Edo Marketplace

Kitauemon's experience was far from unique. The efforts of Edo-period lumbermen to prevent or remedy problems such as his produced an elaborate provisioning system. The wood market in the city of Edo exemplified the system's organization, procedures, long-term development, and problems, even though Edo absorbed more lumber, from more distant suppliers, than did other markets of the day.

THE EARLY EDO MARKETPLACE

Before 1590 a few unspecialized purveyors of lumber and other goods were already established in the vicinity of Edo, a small fishing village at the foot of a rickety little castle, and Asakusa, a minor temple town about five kilometers away on the Sumida River. That year Tokugawa Ieyasu transferred his headquarters

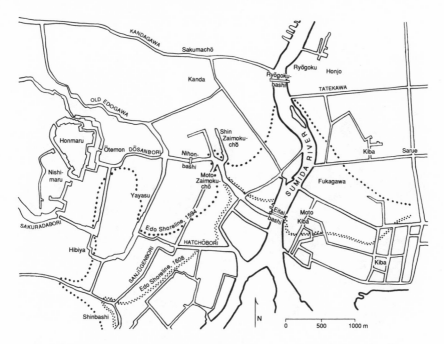

Map 6. **Downtown Edo.** Based on *Tōkyō Kinbō chūbu* (Tokyo: Rikuchi sokuryōbu, 1880), a single-sheet map.

from Sunpu to Edo, and during the next two decades he encouraged merchants who had served him in earlier years to come and assist in his great castle and city building projects.[28] Working as quartermasters, these *goyō shōnin* arranged for loggers and shippers to furnish the goods required by government builders.[29] When major phases of castle work were completed, as in 1606 and 1611, they were allowed to sell any remaining stocks of lumber.

The lumberyards *(kiba)* were situated along the waterfront, but their actual locations kept changing, especially during the seventeenth century as the central city grew.[30] Initially Ieyasu had lumber stored in the Ōtemon vicinity, and merchants beached logs at landings they carved out of the brackish coastal swamp that lay just east of the castle. The sites are now identified as Shinbashi, Hibiya, and Yayasu (Yaesu). As armies of corvée labor filled the swamp with soil from Kanda hill and city blocks took shape on the dry land between newly created canals, urban fires became more and more common. Following a fire in 1631, city leaders

Figure 5. **A lumberyard in Edo.** Workmen stack shingling and rip timber at a canal-side lumberyard on the Tatekawa Canal at the south edge of Honjo district. Large timbers, heavy planking, shingling, and fuelwood are stacked horizontally, but small poles and smaller sawn goods stand on end. Charcoal is packed in straw sacks and covered against the rain. Other lumberyards are visible on the south (Fukagawa) side of the canal. Reproduced from an unidentified partial edition of the series of prints "Thirty-six Views of Fuji," by Hokusai Katsushika (1760–1849), in the author's possession.

ordered the lumbermen to move their yards farther out of town, and they established new ones at Sakumachō, Moto Zaimokuchō, Shin Zaimokuchō, and along the major canals known as Sanjū-genbori and Hatchōbori. The city continued to grow and burn, however, and after the Meireki fire of 1657, authorities pressed the lumbermen to build on the swampland east of the Sumida. They did so, mainly in Fukagawa, constructing an elaborate maze of canals, bridges, and landings. Eventually they also established yards to the north in the Ryōgoku-Honjo area and as far eastward as Sarue.

During the early seventeenth century *goyō shōnin* continued to handle wood they received from *bakufu* and *han* authorities as well as wood they obtained through their own dealings with loggers and shippers. As government timber needs declined, however, they functioned increasingly as marketing agents for independent providers, advising them on market needs and handling their goods for a commission. In the process they evolved into professional wholesalers, specializing by both geographical source of supply and type of commodity. Ever more involved in a nearly unregulated commercial marketplace, they developed informal trade associations to stabilize the market by regulating both its membership and its commercial procedures.

Shortly after New Year's Day in 1657 the infamous Meireki fire ravaged Edo, creating a huge demand for lumber. With government encouragement, entrepreneurs and *bakufu* and *han* officials inundated the city with timber, much of it inferior in quality because previous logging had taken so many of the best stands. The wholesalers tried to handle this explosion of market activity through their existing trade arrangements, with the avowed aim of maintaining price stability and quality standards. Doubtless their efforts also protected their own profits, however, and they were loudly denounced by rival entrepreneurs who hoped to exploit the suddenly expanded market. *Bakufu* leaders also complained, accusing the wholesalers of selfishly restricting access to the trade and thereby hindering urban reconstruction. Ordered to stop obstructing the sale of timber, the wholesalers capitulated. Whether the resulting unregulated market activity expedited Edo's reconstruction is unclear, but demand proved so great that both established and new merchants flourished for several years.[31]

During the 1660s, as the building boom passed, lumbermen again faced hard times, which led to a plethora of lawsuits by and against them and to renewed attempts at market management. Even rural lumber merchants felt the recession: in 1668 a lumberman in the Kinugawa area of the northern Kantō wrote that he had decided to stop sending timber to Edo because the glut of wood had depressed prices so much.[32] The market turmoil and legal wrangling finally led *bakufu* officials to abandon their insistence on unrestricted trade in hopes that registered merchants would provide more orderly marketing and stable pricing. With that shift in official attitude, further structuring of Edo lumber marketing became possible.[33]

During the 1660s and 1670s the processes of timber buying and selling became more clearly separated. Increasingly the guilds of city lumber wholesalers, with their canal-side timber landings and lumberyards, dealt with suppliers while the guilds of brokers, who were essentially bankrollers and intermediaries, handled the sales to retailers, master carpenters, or government buyers.[34] The distinction between wholesaler and broker was still incomplete, and their guilds were not officially recognized, but a few decades later, in 1721, they acquired legal status when the *bakufu* acknowledged their responsibility for marketing timber in Edo. Although the edict of that year neither fixed the number of lumber merchants nor forbade others to enter the business, it instructed lumbermen not to raise prices after fires and not to sell frivolous merchandise. It also ordered them to report the name of anyone who joined one of their guilds or started his own lumber business or anyone who started marketing unauthorized goods.[35]

In following decades the Edo lumber market became more tightly regulated, with the number of recognized merchants fixed, the distinction between *ton'ya* and *nakagai* clarified, their procedures more elaborately regulated, and their formal urban duties and obligations specified. By 1750 these changes had transformed the unofficial guilds of a century earlier into carefully regulated, official guilds or *nakama* whose seats *(kabu)* merchants could buy, sell, and lease as business conditions dictated. It was those arrangements, as noted earlier, that so irritated ambitious young Noguchi Shōzaburō in the 1830s.

WHOLESALERS AND BROKERS

Wholesalers and brokers thus became elaborately organized as regulation modified market forces in the city lumber business. By 1810 the wholesalers were consolidated into three major guilds: the Fukagawa Kiba, combined Ita-Kumano, and Kawabe lumber wholesalers' *kabu nakama*.[36] The brokers, meanwhile, had organized themselves into three groups on the basis of their locations, being known in consequence as the Gokasho, Nanakasho, and Kyūkasho nakagai kumiai ("five-site," "seven-site," and "nine-site brokers' association").

In the early days of Tokugawa rule lumbermen from the Tōkai region had played key roles in building Edo, and they were the major figures in what became the Fukagawa Kiba and Ita-Kumano guilds. As years passed, membership in those guilds swelled and shrank erratically, but they commonly numbered about a dozen or two apiece, and guild members continued to concentrate on handling Tōkai-Kinki timber that came by boat. The ships that brought Tenryū wood from Kakezuka, for example, usually had a capacity of three to eight hundred maritime *koku* (approximately three to eight thousand cubic feet) and were operated by a captain, bursar, cook, and several crewmen. From Kakezuka the captain sailed his laden vessel across Suruga Bay to Shimoda (but to Uraga in later decades), where maritime inspectors from the local magistate's office examined the ship to ensure that it carried neither weapons nor unauthorized money. After receiving clearance, the crew sailed their vessel on to Edo, anchoring offshore at Fukagawa. There they transferred the wood to sampans or dumped it into the water, where workmen formed it into rafts. These they punted along authorized canals to the lumberyard, where the wholesaler officially accepted the goods. Then brokers arranged their sale to the consumer.[37]

Members of the Kawabe (literally "riverside") guild handled raftloads of wood from the mountainous rim of the Kantō, receiving most of them via the old Tone, Ara, and Tama Rivers. On the Ara, for example, lumbermen from Naguri sent their rafts, laden with charcoal and other goods, on their five-day trip down the Naguri to the Iruma, past the government tax station, and thence via the Ara and Sumida Rivers to Senju, where the wood was consigned to the wholesaler's representative. From there the Sumida bore it to the wholesale yard, whence it was marketed by a broker.[38] In the early days these Kawabe lumbermen, lacking old ties to Ieyasu, were inactive in government provisioning and were of little consequence. As reclamation and riparian work opened the Kantō periphery to exploitation, however, and as commercial demand replaced government lumber orders, the Kawabe guild outpaced the other two, by the mideighteenth century numbering some five hundred member firms and dominating the Edo trade.[39]

The individual wholesalers who handled this incoming lumber belonged to specific guilds, but in practice they tended to move from one to another as the ebb and flow of business dictated, giving the system more flexibility—but also more internal disorder—than its formal organization suggests.[40] To minimize competition and discord among themselves, the guilds not only acquired their goods from different regions but also had their subgroups *(kumi)* deal with sellers from designated areas, such as Ōme or Nikkō. In addition the individual wholesalers specialized in terms of commodities, dealing primarily in heavy timber, long sawn goods, poles, or cooperage and roofing. In return for government recognition of their operating licenses, they performed specific public services, most notably canal dredging, bridge repair, and coast guard boat and dock maintenance.

The brokers, who eventually numbered some 550 firms in all, functioned somewhat differently. Dealing with retailers and other buyers in the city, they did not specialize in terms of product and were grouped by location in the city rather than by source of supply.[41] Like the wholesalers, however, they used licensing arrangements to control their membership and regulate activity, and just as the Kawabe group became predominant among wholesalers, so the Gokasho nakagai kumiai became the leading brokers' association.[42]

The segregating of buying and selling eliminated direct competition between wholesalers and brokers, but it scarcely solved all their problems. Difficulty in financing the business persisted— indeed, it intensified. From the 1660s onward, delinquent brokers were a chronic and worsening problem for wholesalers, and in the eighteenth century basic regulations of wholesalers' guilds admonished members not to deal with brokers who were in arrears. Any member who found himself saddled with a nonpaying broker was required to meet with fellow guildsmen to discuss ways of handling the problem.

In addressing the problem of delinquent brokers, wholesalers were constrained by two facts: they relied on brokers to dispose of their goods, and the bidding technique of marketing reduced the wholesalers' market exposure by making the broker a key speculator and risk taker. This problem of market exposure had worsened as time passed. Originally wholesalers handled timber on commission, which they fattened by adding service charges to their bills. They gradually found themselves having to bury those charges in their commissions, however, and they also had to guarantee minimal sales prices and make greater prepayments to the rural lumbermen with whom they dealt. By the nineteenth century they might be advancing as much as 70–80 percent of the anticipated sale price, in effect transforming themselves from commission agents into buyers.[43] Under pressure to move their goods and recover their costs, the wholesalers sought to keep even indigent brokers afloat. One strategy for collectively managing a delinquent broker was to collect a portion of the debt and reschedule the rest for repayment with interest over three years or so. Should the debtor fail to meet the revised schedule or be particularly difficult in negotiations, his name was publicized and guild members boycotted his shop until he made good his debts.[44]

These wholesaler-broker arrangements, which lasted until 1868 save for a brief hiatus in the 1840s, worked no more perfectly than other instruments of commercial regulation. The mix of stresses and temptations that merchants faced continually led individuals not only to move from guild to guild but to break their own guild regulations. They also spurred brokers repeatedly to bypass wholesalers and deal directly with rural timber merchants. More rarely did wholesalers bypass brokers and sell directly to consumers, though the latter on occasion tried to buy directly. Such problems notwithstanding, the arrangements as a whole survived, and they sufficed to move lumber from shipper to consumer within the city.

In short, entrepreneurial lumbermen met investment requirements and overcame fluctuations of the Edo marketplace by developing elaborate organizations and techniques for coordinating and financing their activities. In some particulars, and in overall scale, the practices of these lumbermen differed from those elsewhere in Japan, but in basics they agreed with the solutions arrived at throughout the country. For two centuries following the construction boom of 1580–1650, this shared understanding of entrepreneurial technique maintained a nationwide lumbering system that met tolerably well Japan's urban timber needs.

Recapitulation and Analysis

Command lumbering was dominant during the construction boom of the early Tokugawa decades, but by the mid-1600s it was giving way to entrepreneurial lumbering, which played the key role in maintaining cities thereafter. This shift occurred because government withdrawal from lumber provisioning left the arena to merchants. Governments reduced their lumbering activity as they completed monumental projects, generally before the 1640s, but the trend was encouraged, no doubt, by the economic squeeze resulting from ever more severe budgetary constraints and timber scarcity.

Most timber-bearing land was under government control, but the sharp reduction in monument construction did not end the opportunities for lumber provisioning because the budget shortages that discouraged monument building also prompted rulers to treat their forests as a source of revenue. Lacking capital to finance lumbering projects themselves, many relied on entrepreneurs to convert wood into wealth. This conversion was possible because fire and decay sustained a high if irregular urban demand. Entrepreneurs attempted to satisfy that demand through the sale of both government timber and the harvest from village and household woodlots.

One effect of the trend to entrepreneurial lumbering was to weaken state control of business conditions, and the expanded role of merchants gave market forces freer play all along the provisioning route from stump to street. By the eighteenth century the resulting economic instability led to creation of elaborate monetary and regulatory arrangements and an industrial infrastructure that linked woodlot to urban marketplace in an integrated network of transactions.

This network facilitated cooperation by people at all points from hinterland via provisioning route to city or town. The several participants handled discrete segments of the provisioning operation and worked out institutional mechanisms and techniques of capitalization to link their segments, overcome economic obstacles, and implement their multistage, multiparty transactions. This system enabled merchant oligopolists, whether village landowners or urban magnates, to replace the state in regulating the overall system and distributing its profits. In the process they became the dominant element in the lumber provisioning work of the later Edo period.

THE PROCESS
OF TIMBER
3 | # TRANSPORT

THE ENTREPRENEURIAL lumbermen who became dominant figures in early modern wood provisioning employed the same methods of timber transportation as did government lumbermen. Theirs was an inherited technology that had a venerable history, but they modified it as needed to meet the changing circumstances of their day.

Legacy and Circumstances

During the millennium before 1600 Japanese loggers provided the timber for great temples, shrines, palaces, mansions, and the lesser dwellings of city folk. As long as the timber supply permitted, they operated mainly within the Kinai basin, where the principal cities lay, but eventually they had to obtain timber from more distant sources. Woodland depletion accelerated sharply from about 1570, leading in a few decades to sharply altered forest conditions that combined with changing social circumstances to produce changes in the mechanics of timber transport.

THE LEGACY

During the seventh to ninth centuries rulers built a plethora of great structures in the Nara-Kyoto vicinity, getting their timber from forests of the Kinai plain and the near slopes of surrounding mountain ranges. About A.D. 745, to cite the best-documented instance, the government of Emperor Shōmu began building the

great Tōdaiji in Nara. Officials located sufficient timber in the
mountains of south Ōmi to erect the sprawling set of temple
buildings. Carters hauled the logs to Lake Biwa, where local
workmen formed them into rafts and poled them south to the Seta
River. From there crews of five or six skilled raftsmen steered
them through the gorges down to Uji, whence they floated to the
mouth of the Kizu, to be hauled up that stream to the landing
(kizu) at Izumi. There they were beached, loaded onto carts, and
hauled overland to Nara for processing and use. Temple officials
directed the work, assigning subordinates to supervise it. They
also directed timber cutting in the forests, while at the landings
officials known as raft masters *(ikadashi)* managed the transport
operations.[1]

A century later, when the capital had been moved to Heian
(present-day central Kyoto), much of the city's timber came from
the forests of mountainous Tanba to the northwest. Work crews
floated logs down the Ōi River to a landing at Umezu, where they
were beached, loaded onto carts, and hauled across the lowland
to lumberyards in the city. There officials marketed them, and
laborers hauled them to building sites where woodworkers processed them into construction pieces.[2]

By the tenth century the lumber demands of the capital were
pushing loggers beyond the nearby forests into adjacent provinces
and across the Inland Sea to eastern Shikoku.[3] As centuries
passed, timber came to the metropolitan area from ever farther
afield. In the 1180s, after civil war had destroyed the Tōdaiji, its
gigantic main building was rebuilt but on a smaller scale, apparently because appropriate timber was scarce. After a considerable
search, accessible trees large enough to provide the main pillars
were located at the western end of Honshu, some thirty kilometers up the shallow Saba River. Their extraction required great
crews of men who leveled the land to form skidways and then
dredged and dammed the river to float the huge timbers to the
coast. From there shippers moved them via boat and raft through
the Inland Sea and up the Yodo and Kizu Rivers to the landing at
Izumi, where carts drawn by 120 oxen apiece hauled them to
Nara.[4]

Records of the centuries from 1200 to 1600 show lumber for
use by governments, temple builders, and merchants being
brought to Kyoto by raft and ship from Shikoku and western

Honshu and by cart and raft from the Kiso River area. Because sea transport around the Kii peninsula was treacherous, wood from Kiso was hauled overland by horse cart.[5] In this Kiso lumbering laborers moved the felled pieces to river's edge by hand or on hand-powered sledges and dropped them into the water to float freely. Downstream, where river flow was sufficient, other workmen snared them and bound them into rafts, which they guided to a landing at Sunomata on the Nagara River.[6] There laborers loaded the sticks onto carts for transport to Lake Biwa and thence to the city. By the fourteenth century a bamboo boom was strung across the Kiso at Nishikori, a landing some fifty kilometers upstream from Sunomata. It stopped the loose-floating logs, which workmen formed into rafts or hauled ashore and stored until receiving an order for their disposition.[7]

In these lumber transport practices one can see the general range of techniques used throughout the early modern period. Timber providers employed winches, sledges, chutes, trestles, river booms, and storage areas. They free floated and rafted logs and shipped them by river and sea both aboard vessels and strapped alongside them. The centuries after 1600 would witness an impressive elaboration of these techniques and their application on an unprecedented scale, but they would see no fundamental technological or conceptual breakthroughs in the field of timber transport. Also, whereas these earlier lumbermen used massive, two-wheeled carts drawn by horses or scores of oxen to haul timber overland, early modern provisioners did not. Indeed, the Tokugawa rulers forbade highway cartage, presumably because vehicles rutted roads and obstructed the way, which they wanted smooth and clear for passage of their own runners, retainers, and retinues. In consequence waterways became the major arteries of timber transport.

THE CHANGING CIRCUMSTANCES

The changes in transportation technique that did occur reflected the changing circumstances of early modern lumber provisioning. The earlier centuries of logging had depleted timber stands in the Kinai region, but elsewhere luxuriant, harvestable forests remained largely intact. From the 1570s, however, that situation began to change as the urban timber provisioning described in chapter 2 started consuming forests from southern Kyushu to northern Honshu. As a consequence, whereas earlier generations of loggers had been able to fell streamside timber, float the pieces to landings, and carry them by ship or cart to the construction site, as the seventeenth century advanced, the last of the easily accessible old-growth timber was cut and loggers moved ever deeper into interior mountains in search of satisfactory wood. This change in forest condition presented lumbermen with a severe challenge and elicited noteworthy engineering and managerial solutions.

Two basic factors made the loss of readily accessible timber so troublesome. First, the topography of Japan is unfavorable to logging. The archipelago lacks tectonic flatlands, consisting instead of slender deposition plains and innumerable V-shaped valleys sandwiched between convoluted arcs and nodes of acutely upthrust mountain range. Out of these mountains tumble streams whose levels of water and debris fluctuate sharply and quickly. Only in their lower reaches do they make serviceable arteries of commerce, and even there erratic flow makes them treacherous. Nevertheless, the streams were essential to early modern loggers because the character of the mountains generally precluded inter-montane transport of all but the smallest pieces, such as cooper-age, charcoal, and fuelwood, which could travel by packhorse. This natural constraint on upland transit, together with the *bakufu*'s prohibition of highway cartage, limited land transport almost entirely to pack animals and coolies.

Second, although rivers were crucial to timber movement, their use was complicated by its social context. Cultivators, as noted in chapter 1, pursued land reclamation energetically during the seventeenth century. As they advanced up narrow valleys and steep hillsides, their work exposed more and more land to erosion and rapid runoff, intensifying the rate of silting and fluctuations in streamflow. Downriver, meanwhile, other cultivators were cre-ating more and more fields out of swamps and flood plains, increasing their exposure to the rivers even as these were becom-ing more erratic and dangerous. The conjunction of these trends necessitated dredging and the creation of ever higher levees to hold streams in their beds.[8] Lumbermen were a menace to these levees and their creators because floating timber could wear away and puncture the levees, flooding nearby villages and fields.

Moreover, tillers wished to put as much arable land as possible into wet-rice production, and many of the irrigation systems that made this possible drew water from streams through elaborate, often fragile systems of dams, dikes, and gates.[9] When logs came hurtling down twisting rivers, they could smash into these structures, to the outrage of affected cultivators. Understandably, downstream villagers often saw upstream loggers as serious enemies, and they acted accordingly, both within and without the law. Loggers could not ignore them.

In short, rivers were essential to early modern timber provisioning, but concurrent trends in agriculture and lumbering were bringing farmers and loggers—often the same people—into conflict in ways that threatened the use of streams in such provisioning. Lumbermen had to deal with that situation if they were to continue furnishing the wood that sustained Tokugawa urban civilization and their own livelihoods. They developed elaborate systems for managing their river use, regulating both its procedures and timing. They also pursued river improvement, clearing streams of obstacles to open more woodland to exploitation and make river use safer and more effective.

River Improvement

During the seventeenth century old-growth stands were devoured in all the great coniferous forests of Japan, most notably those of Shikoku, the Yoshino-Kumano area of Kii peninsula, the Kiso and Tenryū river valleys, and Akita and Tsugaru in northern Honshu.[10] As logging operations moved inland and standing timber became less and less accessible, efforts were made to improve river transport. The process was difficult and the result much less than ideal.

THE PROCESS

The initial objective of river work was to open streams to free-floating logs. Among the most ambitious projects were those of the merchant-offical Suminokura Ryōi (1554–1614). At the behest of Tokugawa Ieyasu he directed riparian work on the Ōi River northwest of Kyoto and later worked on the Tenryū and Fuji Rivers to the east.[11] His crews removed rocks, sandbars, and accumulated debris, making long stretches of river accessible to floating logs and shorter stretches to boats and rafts. Projects of later decades extended the usable length of such rivers. Thus, Suminokura opened some thirty kilometers of the Ōi by 1606, reaching the vicinity of Tonoda village, where the river's main branch turns eastward into the central valley of Yamaguni. Villagers who controlled the rafting completed six additional river-clearing projects between 1628 and 1753 by levying special local taxes to pay work crews of 120–275 laborers who extended the usable length ever deeper into the Ōi's branches and upper reaches.[12]

The process of river opening is nicely illustrated by work done on the Yoshino branch of the upper Ki River.[13] Most of the timber from this watershed went down to Wakayama for local sale or shipment to the lumberyards of Osaka. Throughout the 1610s loggers floated loose pieces down to Gojō town, where raftsmen formed them into rafts for the fifty-five-kilometer run westward to Wakayama. During the next few decades, however, the villagers who handled the logging began a series of river-opening projects designed to replace free floats with rafts. By 1643 they had extended the raft run upstream to Yoshino town, and by 1680 the river was open to rafting as far as Wada Oshima. By then, however, progress had slowed severely: most notably, during the 1660s, four years of slack-season time were consumed cutting through ledges along a few hundred feet of rapids near Kawakami. In following decades the villagers worked at river improvement sporadically, and by the 1750s they finally had pushed up to the village of Shionoha. The gain from those last seventy years of occasional effort was only about ten kilometers, but it brought them at last to the first of the Yoshino's several source brooks.

The last phases of this project were excruciatingly difficult. Scores of men gradually winched great boulders out of the way. In place after place villagers lowered and smoothed outcroppings of ledge. They used hammers and chisels on some. Others they covered with fuelwood, which they soaked with oil and then fired. Burning great quantities of fuel, they gradually heated the rocks to a high temperature, whereupon they flooded them with water. The contraction pressures shattered the ledges, simplifying their removal. This massive investment of time and resources notwithstanding, the water flow at Shionoha was so modest that rafts would not float. Operators had to form temporary splash dams (*teppōen* or "rifle dams"), place a raft on the streambed below one of them, open the gate when the dam was full, and send the raft downstream on the flood.

THE RESULT

Once cleared, streams required regular maintenance, and that work became an inescapable part of the cost of lumber provisioning. Thus on the Tama, raft operators began every logging season by sending downstream three or four rafts bearing labor crews whose task was to locate and clear any new rock pileups, sandbars, or snags that might disrupt subsequent traffic.[14]

For all the trouble rafting entailed, loggers generally preferred it to free floats. Loose pieces, and most especially semiprocessed sticks, could be badly battered on rocks en route to the landing, and they were more likely to be lost in transit. Theft was one problem, sudden storms another; the latter often stranded or buried pieces or washed them out to sea. In 1676, for example, on the Yoshino River in Shikoku, 2,441 or about 12 percent of 20,500 floating pieces were lost en route.[15] Losses on the Tenryū were a chronic problem, and in 1725 log transport there was shifted entirely from free floats to rafting.[16]

Rafting also eliminated much of the damage that lumbering did to levees and irrigation works, thereby reducing the frequency of disputes and lawsuits, which rulers, loggers, and suitors all preferred to avoid. On Shikoku the Yoshino River ran from the interior of Tosa through Tokushima han, and lawsuits over riverine damage led to interdomainal political tensions and, eventually, agreements on raft usage.[17] Local authorities forbade lumbermen to float loose logs on the Tama River, in part to protect a dam and intake structure at Hamura that drew water for Edo into the Tama potable water aqueduct (*jōsui*).[18] In Kaga problems of theft, stranding, and property damage along the river prompted the daimyo as early as 1601 and 1609 to issue special orders on handling logs. Free floating continued, perhaps because river flow or logging operations were insufficient to warrant rafting, but a report of the 1760s regarding a log float on the Kurobe River suggests the ongoing difficulty Kaga faced in reconciling the needs of logging and agriculture.[19]

Pursuant to your report the other day regarding *sugi* sent out from the inner Kurobe mountains, the pieces were put into the river on the twenty-second and in free float reached the landing yesterday, where they were collected. From early next month they are to be sent downstream. However, due to the low water, here and there pieces stop at irrigation inlets. A notice went out a few days ago, and guards quickly shut off the water flow [at the inlets]. If things remain as they are now [i.e., if the water level holds up], we want to send the wood along little by little. Be sure that people place netting across irrigation inlets and brace them with bamboo, and keep the guards alert.

Even rafting could do damage, however. On the Tama, raft maneuvering was so difficult (in part because raftsmen loaded their craft with high-priced produce for sale in the city) that the raft operators' association imposed a regular fee on each raft, the income to compensate villagers for any damage to irrigation inlets along the way.[20] On the Ōi, even after rafts replaced free floats, the daimyo of Kameyama han imposed a regular raft tax to cover the costs of dike maintenance and repair.[21] In Shinano, when the daimyo headquartered at Matsumoto contracted with merchants to get out logs via the Sai River or its tributaries, he dealt with the issue by stipulating that the entrepreneur must settle any village disputes stemming from his logging work.[22]

Because of the problems associated with the movement of logs and rafts, an autumn-to-spring rafting season became established on most of Japan's rivers. The season varied in its timing, depending on local agricultural needs and the rhythm of stream flow, but nearly everywhere rafting was prohibited from late spring until the maturing rice no longer required irrigation. On such major logging rivers as the Tenryū, Kiso, and Ōi near Kyoto, there were official opening and closing dates for raft work. Midwinter stream flow commonly was so modest, however, that rafting was difficult, and loggers did most of their shipping during the autumn and spring months. Thus on the Naguri River low water in winter led raftsmen to send the bulk of their goods in spring and fall, and even then the five-day raft run could be appreciably disrupted by high or low water.[23]

River improvement thus made rafting widely feasible, but the basic character of Japan's geography and agriculture imposed sharp constraints on raft usage. The organization and technology of early modern lumber transport reflected this combination of capability and constraint.

An Overview of Lumber Transport

During the decades in which loggers were coping with forest depletion by improving river flow, higher authorities were developing supervisory systems to control the extraction of timber. The systems were designed to manage the felling, transport, and disposal of diverse goods that belonged to diverse owners and that went to market by various routes and in various stages of processing.

When the great Kiso forest came under the management of Owari han after 1615, for example, Owari officialdom gradually erected an administrative structure that controlled the forests, their exploitation, and the disposal of their yield.[24] The heart of their system was its control of lumber transportation. As developed by the 1660s, three minor *han* officials titled Kiso timber overseers (*zaimoku bugyō*) supervised forest management, logging, and the transport of timber from mountainside to the boom and storage area at Nishikori. At Nishikori three superintendents (*bugyō*) assumed rotating responsibility for operation of the boom and storage, maintenance of the river, collection of transit taxes and user fees, and the dispatch of rafts to the lumber yards at Shiratori. At Shiratori two timber overseers, one on duty at any moment, were charged with storing and disposing of the lumber. The overseers at both sites had support staffs. That at Nishikori totaled some seventy-five to eighty men, consisting of comptrollers, investigators, clerks, guards, river maintenance supervisors, and their assistants. This staff in turn oversaw the work of the professionals and labor crews who actually constructed and installed the log boom and beached, sorted, measured, stacked, and rafted the lumber.[25]

The arrangements in other timber-producing areas differed in particulars. Nearly everywhere, however, the rulers controlled the bulk of timber-yielding forestland and regulated the transport of lumber. In part they did so to prevent or resolve disputes over river use (e.g., irrigation vs. log floating; fishing vs. rafting). In

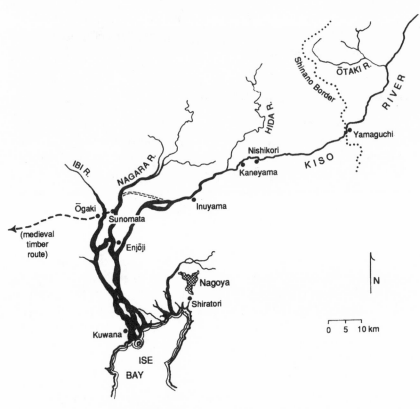

Map 7. **The Kiso River and Region.** Adapted from Tokoro Mitsuo, *Kinsei ringyōshi no kenkyū* (Tokyo: Yoshikawa kōbunkan, 1980), 2.

part they were assuring their own access to wood for castles and other monuments to their glory, and in part they were interested in the income generated through sales, taxes, and user fees.

Although the rulers oversaw timber transport, only a portion of what traversed the rivers was theirs. Some wood belonged to entrepreneurs, whether rural or urban, the proportion varying greatly from river to river. On the Tama almost all timber was commercial; on the Yoneshiro almost all was governmental. On the Ōi west of Mt. Fuji the two were thoroughly mixed. But everywhere ownership was recognized, with free-floating logs bearing brands or stamp marks and rafts displaying flags or other insignia. At various points in transit, moreover, workmen sorted the lumber by owner to assure that its disposition conformed with prior contractual arrangements and local practice.

Only a fraction of the wood that went downstream was in the shape of logs. Woodsmen processed a substantial portion into semifinished form at the felling site or at convenient places en route to market. They could do so because the general dimensions of the intended lumber were already known, either thanks to the felling order, which specified number, type, and size of pieces to be provided or because the yield was of a customary type, such as cooperage, roofing, or standard-size housing posts and beams. Several other factors encouraged processing at an early stage. It left the wood easier to transport and eliminated waste in shipping—a substantial saving to the lumber provider. Also, the necessary skilled labor was available in the woods, where it probably was cheaper than in city or town because living conditions were more primitive and alternative employment opportunities few.

A rich variety of wood products, known by a dazzling array of local names, went downstream to market. They can be categorized roughly as (1) regular logs *(maruta)*, commonly twelve or eighteen feet (two or three *ken*) in length; (2) squared pieces *(kakuzai)*, split or hewn by broadaxe, most commonly to about four, six, or eight inches *(sun)* on a side; (3) split pieces (known as *kureki, doi,* and by many other names), mostly destined for cooperage or shingles;[26] (4) flat pieces, equivalent to boards and planking, whose names and dimensions varied greatly; and (5) fuelwood, mostly broadleaf growth usually known as *zatsuboku (zōki)* or "miscellaneous wood," much of which was converted to charcoal before shipping.

Most of these items went to market by river, but some went overland. Thus, cooperage from the Chikuma-Shinano river valley went overland by packhorse because the river flowed northward, away from major markets, creating a prohibitively long supply line, and because a geologic fault downstream created a rapids so violent that rafting was impossible and free-floating wood, especially semifinished pieces, could be so severely damaged as to preclude the river's use.[27] Similarly, sending wood downstream from the upper reaches of the Ōi River west of Mt. Fuji was so difficult that loggers commonly portaged the goods over ridges to other rivers for shipment.[28]

Such anomalies notwithstanding, almost all forest products, including processed shingles, went by raft. The rafts were complex instruments, their size, configuration, and mode of construction differing with the locality and the material being shipped. Mostly they were stiffened by cross pieces and bound together by vines, and often they carried some other material, whether smaller pieces of lumber, fuelwood, charcoal, or other local produce. In the upper reaches of rivers the rafts were commonly one or two timbers in length and four to ten feet *(shaku)* wide. Downstream they were bound into wider and longer rafts, and in lower stretches they might be several hundred feet in length, steered by as many as eight crewmen, although rafts of 80 to 180 feet were more common.[29]

Kiso Lumber Transport: From Logging Site to Boom

The physical process of getting timber from mountain to metropolis consisted of several stages, the particulars varying from river to river.[30] The well-studied lumber transport system of the Kiso River, as managed by Owari han, illustrates the process. Its stages can be identified in this way:

1. assembling the logs[31]
2. working logs down the mountainside *(yamaotoshi)*
3. sending logs down the ravine *(shokokugari)*
4. floating logs down the river *(ōkawagari)*
5. processing logs at the boom
6. rafting logs to the shipping point
7. marketing and shipping timber by sea

In Kiso, as in most mountainous areas, logging was a valued source of income for villagers, most of whom had little or no agricultural land. Logging offered attractive part-time employment because felling and shipping were mostly winter tasks, coming when other jobs were few. They were winter work for several reasons. When deciduous trees were bare, the sap down, river flow relatively stable, and—in colder areas—the ground frozen or snow covered, wood could be taken from the forest with the least amount of effort or damage. Also, farmers did not use irrigation water then, so loggers could move timber with relatively few social complications.

ASSEMBLING THE LOGS

At a logging site professional choppers felled, limbed, and cut to length the trees that supervisors had marked. Depending on the

work order, the terrain, and the trees being felled, skilled woodsmen might also square, split, or saw the logs to shape there. Work
crews then moved the sticks to a nearby assembly point, either
dragging them on the ground or moving them on small sleds. In
this work they used long-handled hooking tools *(tobiguchi)*, rope
made of hemp or heavy vine, and winches *(rokuro)* when the timbers were especially large or the terrain perverse.

Sled design varied with the situation. Sledges for use on bare
ground were about one and a half feet wide and had runners five
to eight feet long. They were made of lighter-weight fir except for
the runner shoes, which were of oak or other hardwood that
withstood the wear and tear of use. Workmen loaded some three
koku (thirty cubic feet) of wood onto their sledges and moved
them on rollers or dragged and pushed them along rough roads or
skidways made of brush. In deep snow they used sleds with ski-
like runners about four inches wide. Sleds used on steeper slopes
(over 10 percent grade) had four-foot runners and a single cross-

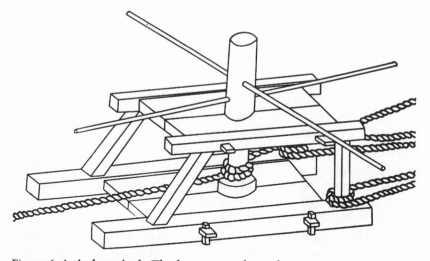

Figure 6. **A timber winch.** The *kagurasan,* shown here, was a common type
of timber winch or *rokuro.* Ropes looped about the legs of the winch and
exiting right anchor it to trees to hold it in place. The rope that drags, lifts,
or lowers the timber being moved is wrapped two turns about the capstan
to provide a friction grip and held snug by a man standing offstage at right
rear. Four workmen move the log by turning the capstan, using the project-
ing arms, which are drawn too slender in this illustration. Reproduced
from Tokoro Mitsuo, *The Wood-Cutting and Transporting System of Kiso*
(Tokyo: Tokugawa Institute for the History of Forestry, 1977), 114. Courtesy of the Tokugawa Institute and its director, Ōishi Shinzaburō.

timber (bunk). Workers lashed the front ends of logs to the bunk, letting the rear ends drag in the snow to brake the descent. On more level ground they used double-bunk sleds with runners six or seven feet long, and because those moved more easily on snow, they could carry five to seven *koku* of wood per load.

Labor efficiency varied greatly with conditions, but a sled crew could move about three loads of wood five to six kilometers a day in reasonably suitable terrain. The assembly point that was their destination might be the place where pieces were dumped into the river, a central storage point, or a spot where logs were hewn, split, or sawn into lumber.

WORKING LOGS DOWN THE MOUNTAINSIDE *(Yamaotoshi)*

From the assembly point wood left the mountains in inchworm fashion. Workmen used the logs being moved to form elaborate timber chutes whose character depended on the terrain. They might be three-foot-wide trestlelike frames *(sade)* built on scaffolding that elevated them enough to achieve a reasonably consistent incline. Or they might be trough-shaped chutes *(shura)* of parallel logs that rested directly on the ground or on horizontal bridging.

Commonly a lead crew *(kihana)* of eight or sixteen men (one or two labor gangs or *kumi*) laid out the chute, building it downhill length by length for a total of 80 to 250 feet. At its foot they arranged other pieces to form a log catchment. Enough additional logs were then chuted down from the assembly point to make a dispatch of 100 to 150 logs. After their descent a cleanup crew *(kijiri)* of eight or sixteen men followed, disassembling the chute and sending on the logs as it advanced. At the catchment basin a third crew of about eight men received the incoming logs and moved them on to the lead crew, which was by then busily laying out the next segment of chute and catchment. This was dangerous work, and it involved much calling back and forth as those below instructed those above when to wait and when to send down more pieces. At corners or other troublesome places along the chute, workmen were posted to keep the traffic flowing and to reposition any pieces of chute that became dislodged.

The chutes and trestles generally followed the steep valleys carved by rushing streams, and the task of erecting them over and around ledges, among boulders, and down precipitous slopes

Figure 7. **A trestle descends to streamside.** By careful planning and construction, workmen in Kiso created tortuous but smoothly descending,
partially elevated skidways for their logs. Here scarcely visible workers at
lower left use pole-hooks to slide pieces down the trestle *(sade)* to the
river, while others steer their float to Nishikori. Reproduced from Tokoro
Mitsuo, *The Wood-Cutting and Transporting System of Kiso* (Tokyo:
Tokugawa Institute for the History of Forestry, 1977), scroll 1, plate 12.
Courtesy of the Tokugawa Institute and its director, Ōishi Shinzaburō.

made *yamaotoshi* the most difficult stage of the transportation
process. Laborers might employ winches to haul large pieces out
of depressions or lower them over precipices; they regularly used
the *tobiguchi* to snake pieces along trestles. Where logs slid too
rapidly, workers slowed them by suspending over the chute curtain logs *(noren)* that descending logs had to push up and slide
under.

The rate of progress varied greatly with the terrain. On a

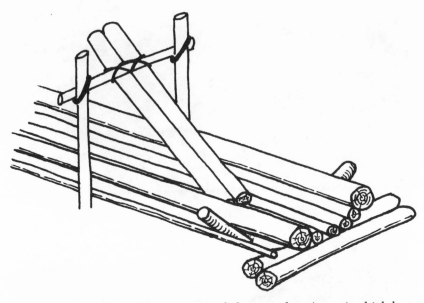

Figure 8. **Curtain logs.** The two suspended curtain logs *(noren)*, which logs
had to push up and slide under, controlled the speed of wood descending
the chutes *(shura)* erected on steep hillsides. Derived from Nihon Gakushiin, *Meiji zen Nihon ringyō gijutsu hattatsushi* (Tokyo: Nōkan kagaku
igaku kenkyū shiryōkan, 1980), 392.

rough average, each worker at a catchment could move about 40
to 60 logs a day. Eight men could thus presumably move 300 to
500. Catchments commonly held 100 to 150 logs. It seems to follow, therefore, that in one day a complete *yamaotoshi* crew consisting of twenty-four to forty men would be able to move 100 to
150 logs through two to four chute-catchment segments—a total
distance that might be as little as 150 feet or as much as 1,000.

SENDING LOGS DOWN THE RAVINE *(Shokokugari)*

The third stage of log transportation commenced at the point
where stream flow was sufficient to serve the transporters. Similar
in character to *yamaotoshi,* it differed mainly in that it was a
larger-scale operation and made much greater use of water power.
Larger crews worked larger numbers of logs along, stage by stage,
relying primarily on stream flow, which they channeled and regulated to maximize its usefulness.

In place of catchments, men built splash dams to form holding pools. The lead crew used a batch of logs, about two hundred

koku in volume, to construct the dam. Another six hundred to one thousand *koku* of wood accumulated in the holding pool, where a crew of four men aligned it for passage through the gate. When the water level at the dam was high enough, the gate was knocked loose and pieces floated out on the torrent. As they moved downstream, the lead crew placed them along the streambed to improve stream flow. They lodged some near ledges and rocks to reduce stranding and damage to floating logs, and they

Figure 9. **Splash dam and chute.** To bypass ledge outcroppings or boulders along small streams, workmen erected temporary structures of this sort and then used pole-hooks and water power to steer timbers along them. Reproduced from Tokoro Mitsuo, *The Wood-Cutting and Transporting System of Kiso* (Tokyo: Tokugawa Institute for the History of Forestry, 1977), scroll 1, plate 15. Courtesy of the Tokugawa Institute and its director, Ōishi Shinzaburō.

employed others, together with rock and gravel, to raise the water level by narrowing the channel at wider places. In spots this construction work amounted to the creation of elaborate *shura,* down which water cascaded, carrying the logs along. After all the logs in the holding pool had been deployed along the next length of streambed and splash dam, the cleanup crew disassembled the original dam and worked its pieces, and after that the pieces that had been laid out along the streambed, on downstream to the next holding pool.

The distance between dams varied with the situation, but in general this process of working some eight hundred to twelve hundred *koku* of logs downstream required crews totaling sixty-four to eighty-eight men. They would set up, use, and tear down four or five dams per day, tasks that might move the wood anywhere from a fraction of a kilometer to as much as two kilometers along the river.

FLOATING LOGS DOWN THE RIVER *(Ōkawagari)*

The processes of *shokokugari* and *ōkawagari* differed more in degree than in kind. The latter was an expanded version of the former, occurring on larger rivers and covering longer distances, but requiring fewer workmen. Lead crews and cleanup crews worked larger quantities of free-floating logs downstream, usually handling twenty thousand to thirty thousand *koku* per float. To move such a volume from the confluence of the Ōtaki and Kiso Rivers to Yamaguchi on the Shinano border required a total crew of thirty-five; from Yamaguchi to Nishikori, about thirty men. The combined distance was eighty kilometers; the number of days required for a log to cover the distance is unclear, but likely was about seven.[32]

The work of river preparation in *ōkawagari* was essentially similar to that of *shokokugari,* but it was much less concentrated. If necessary, workmen built splash dams and narrowed the river artificially to enhance water flow. They set up guide poles and ropes and even protective barriers of beached or anchored logs at treacherous spots and bends in the river where logs might be damaged or stranded or might snag and create jams. Where river flow was too rapid, they might erect barrier ropes, rafts, or dams to slow or halt the logs temporarily, enabling crews to control their movement more closely.

Figure 10. **Workmen guide free-floating wood.** Using their pole-hooks, men steer some logs to form protective barriers around the rocky outcroppings and dislodge and steer others down the fast-flowing river. To reach the rocks in midstream, the men rode small rafts of the sort shown in the lee of the ledge at center. Reproduced from Tokoro Mitsuo, *The Wood-Cutting and Transporting System of Kiso* (Tokyo: Tokugawa Institute for the History of Forestry, 1977), scroll 2, plate 3. Courtesy of the Tokugawa Institute and its director, Ōishi Shinzaburō.

Despite the preventive work done by lead crews preparing the river, as floating logs moved from the gorges of Shinano out onto the broader flatlands of Mino Province, they sometimes damaged levees, riverbanks, bridges, irrigation works, or other riparian structures. To minimize such damage, *han* authorities restricted floats on the Kiso to about ninety days during the winter and suspended them during high water, when the risk of both damage and loss was greatest. Moreover, the government maintained way stations *(bansho)* at regular intervals, and officials stationed there oversaw the float and dealt with any problems that arose.[33] These officials were under control of the Kiso timber overseer, but when the logs floated into the still water of the Nishikori boom and landing, they came under the Nishikori superintendent's jurisdiction.

Kiso Lumber Transport: From Boom to Market

At Nishikori, logs drifted quietly along on the placid water past a towering rock pillar that jutted out of the river some thirty feet offshore to the right.[34] As the logs slowly swept around the rock, they bumped up against the boom, a woven vine-and-log apparatus that stretched from the rock diagonally downstream across the river to a cluster of stakes on the opposite bank, several hundred feet away.[35]

At the boom, men sorted the pieces of wood, separating them

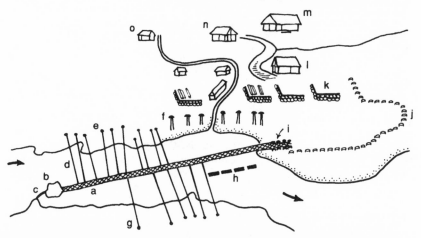

Figure 11. **The Nishikori boom site.** Here free-floating logs collected behind the boom. Workmen pushed them under the boom and assembled them into rafts for the run to Shiratori or stored them on the overflow area and platforms until they could be sent along. Key: (a) the boom, (b) anchor rock, (c) boat channel, (d) truss lines to cliff, (e) truss-line stakes, (f) emergency truss-line stakes, (g) stabilizing lines, (h) ballast baskets, (i) anchor stakes, (j) overflow posts, (k) storage platforms, (l) labor rest shelter, (m) business office, (n) guard office, (o) rope storage. Derived from Tokoro Mitsuo, "Nishikori tsunaba ni tsuite—kuchie kaisetsu," *Shakai keizai shigaku* 2, 12 (Mar. 1933): 105.

by owner and type. The yield was measured, appropriate taxes and user fees levied, and the pieces disposed of. Should the boom grow full, the water rise, or the rafting season end, workers pulled the wood ashore and stored it until cleared for rafting onward. Otherwise they slipped pieces one by one under the boom at the processing point *(kakitateguchi)*, turning them over to raftsmen whose practiced hands quickly lashed them together to form rafts for the next stage of the journey. If all went well, eight days later the raftsmen would deliver their wood to Shiratori, about a hundred kilometers to the southwest. From there it went by sea to Edo, Osaka, or several nearer towns and cities.

THE NISHIKORI BOOM

One of the largest and most actively used booms of the Edo period, that at Nishikori, is also the best studied. Originally, in the fifteenth century, it consisted of bamboo poles lashed together. Later, wisteria vine *(fujitsuru)* was used to make a slender rope, and by the 1720s a heavier vine *(shirakuchizuru)* became standard. The hawser grew in diameter from six inches to nine and eventually to one foot, and by the 1790s from a single rope to a complex structure using four of these hawsers and elaborate supplemental paraphernalia. Each of the four hawsers consisted of about seventeen tons of vine cut into eight- to ten-foot lengths and twisted together. Each was used for two seasons, being strung up in the autumn, taken down in early spring, and stored in a shed through the summer.

To form the boom, men stretched the hawsers across the river and anchored them to the great rock at the upstream end and to the cluster of ten stakes downstream. The stakes actually were *hinoki* logs that measured a foot thick and fifteen feet long. The workmen set them nine feet into the ground, packed the earth firmly around them, and overlaid the surrounding surface with rock. After tieing the boom at both ends, they reinforced it with a complex set of attached buoy logs and sticks and anchored it both upstream and down with an array of truss, stabilizing, and ballast lines made of hemp and measuring two to three inches in diameter and up to four hundred feet in length.

The installation of this massive device was a major undertaking. Late in the fall each year, laborers gathered some sixty tons of vine from surrounding areas for new hawsers and stored it to

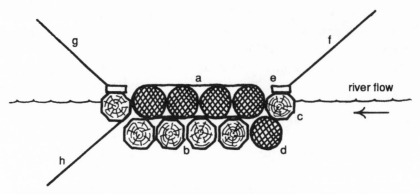

Figure 12. **Cross section of the Nishikori boom.** This sketch shows how the boom's several parts fitted together to provide strength, stiffness, and stability in the face of erratic stream flow and immense lateral pressure. Key: (a) four main hawsers, (b) four buoying logs, (c) buoy logs fore and aft, (d) stabilizing hawser, (e) truss-line anchor stick, (f) truss line to cliff, (g) stabilizing line and anchor stick, (h) ballast line. Derived from Tokoro Mitsuo, "Nishikori tsunaba ni tsuite—kuchie kaisetsu," *Shakai keizai shigaku* 2, 12 (Mar. 1933): 106.

cure. The following spring, twelve expert rope makers labored together for thirty-six days to weave the two new strands. When time for installation came, men hauled these strands, together with two from the previous year, to water's edge and dropped their ends into the river for ten days to soak up water and soften enough for fastening at the rock and stakes. Then, five days before the floating season began, a work crew of 105 to 110 men from nearby villages began the methodical work of constructing the boom.

On the first day they hauled the four hawsers across the chilly river, attached them to the great rock at the far end, and anchored them loosely to the ten *hinoki* stakes at the near end. On the second day the crew floated into place the fifty-six twelve-foot *sawara* logs that gave the boom buoyancy and stiffness. To hold them in place, they attached a fifth hawser, which served as a buffer and stabilizing line on the upstream side of the boom. Then, with the main pieces of the boom in place, they tightened it to the desired tension and bound it firmly to the rock and stakes.

On the third day, the day of "the big binding," crews lashed the four hawsers together at three- or four-foot intervals with heavy vine, which they bound in place with lighter vine. The following day—day four—they strung the eighteen upstream truss lines, fastening them to *sawara* anchor sticks, which attached to the buoy logs lining the near edge of the boom, and running them high above the water to trees or stakes on the cliffs overlooking the boom site. They also attached the several downstream stabilizing lines, which were anchored on shore, and the ballast lines, which anchored to rock-filled bamboo baskets resting on the streambed. With the anchoring complete, the boom was ready for service. Its disassembly in the spring required only half as large a crew.

The boat channel on the far side of the great anchoring rock was closed off with a seven-inch, two-ply rope equipped with its own buoy logs. This could be opened to accommodate river traffic. Some 250 feet upstream was another, simpler hawser that held logs temporarily to expedite the sorting of government from merchant lumber.

The Overflow and Storage Areas

During periods of normal streamflow, logs collected above the boom, where work crews from nearby villages processed them on a regular schedule. If the river rose, however, the entire work force turned out, day or night, to handle the flood of wood. All boat traffic was halted, workers tightened the truss lines, added additional heavier ones as necessary, and laid logs up on the boom to hold it down in the roily water. The accumulating logs might build up three to five deep, numbering as many as thirty thousand. The water level could rise as much as ten feet.

As the river rose, it spread onto the sandy overflow area that lay beyond the anchor stakes. Measuring some two hundred by six hundred feet, the area was bounded on its lower edge by a fence of some eleven hundred posts similar to those anchoring the boom. As the river overflowed, pieces of wood drifted up against the posts and others collected behind them until as many as eleven thousand logs, one layer deep, might be trapped. If the water rose high enough to layer them, the number could swell, as it did in about 1866, to seventy-five thousand pieces. Every autumn before the float commenced, workers dredged the overflow area to bring it down to optimal level, which sometimes meant removing four feet of accumulated sand, gravel, and debris. Every five years they had to replace the overflow posts.

Usually some sticks still remained at the overflow site when the rafting season ended on the "eighty-eighth night" after New Year's, and yardmen stacked them in the adjoining lumberyard until the autumn. The yard, which was situated on higher ground between the overflow area and superintendent's business office, consisted of rectangular platforms surfaced with large, flat rocks. Each platform measured twenty to thirty *tsubo* in area (roughly twenty-five by forty feet) and held twenty stacks of racked wood. If the wood was split pieces, each platform could hold 20,000.[36] The platforms came into use in the 1660s, perhaps because intensified use of the river had led by then to tighter restrictions on the rafting season. During the 1720s their number grew from four to six, accommodating a total of 120,000 split pieces to be held over the summer if necessary.

The volume of wood that floated through Nishikori varied substantially as the decades passed and the condition of the Kiso forests changed. Tokoro Mitsuo reports that boom crews there processed between 180,000 and 400,000 *koku* (1,780,000 to 3,950,000 cubic feet) of lumber per year.

RAFTING TO SHIRATORI

As workmen slipped logs under the boom at the processing point, an official counted and measured them, and two raftsmen bound them together.[37] Normally a Nishikori raft consisted of two lengths of log, each piece twelve feet long and eight to fifteen inches in diameter, enough being bound together to form a raft about ten feet wide. Besides the basic layer of logs, additional pieces were attached to link front and rear sections, to secure crosspieces, to provide added stability, and to give the raftsmen better standing and steering positions. The whole was lashed together with vine. When complete, such a raft might contain up to twenty-five or thirty *koku* of wood, or about fifty logs. The number of rafts prepared daily at Nishikori ranged from 30 to 150, depending on the size and number of pieces available. The maximum allowed was 200, and about 50 per day was average.

When a raft was ready to go, two men boarded, one fore and one aft. Along much of the four-kilometer run to Kaneyama the river was full of rocks and rapids, and the two raftsmen had to steer carefully to avoid breaking up their craft. From the landing at Kaneyama the river was much easier to navigate, so one man

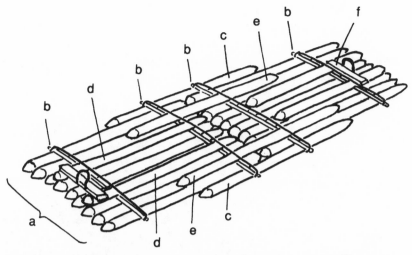

Figure 13. **A Nishikori raft.** In its initial configuration as a small-stream raft, it consisted of two log lengths bound together by vines, crosspieces, and linkage logs. The particulars of raft construction varied from river to river, but sections such as this were lashed together to form the long rafts that snaked down the lower reaches of major rivers. Key: (a) nine-log foundation, (b) stabilizing cross pieces, (c) flanking linkage logs, (d) crosspiece locking logs, (e) locking and linkage logs, (f) raftsman's rudder platform. Derived from Nihon Gakushiin, *Meiji zen Nihon ringyō gijutsu hattatsushi* (Tokyo: Nōkan kagaku igaku kenkyū shiryōkan, 1980), 429.

disembarked to walk back up to Nishikori for another raft ride the next day. The other rode another twenty kilometers to Inuyama, which he had to reach by nightfall. Upon arrival he obtained a receipt for the delivery and turned the raft over to a crew under the control of officials stationed there.

The next day the Inuyama crew linked two rafts together to form a longer one, and a single man steered it the sixteen kilometers to the Enjōji way station, where he turned it over to a crew under authority of officials there. This crew lashed between thirty and sixty of the basic Nishikori rafts together to form a much longer one.[38] During its fifty-kilometer run to Shiratori, it was escorted by one or two boats and manned by a crew of eight. This last leg required six days of daylight travel. As the raft entered the Kiso delta, its crew steered it down a secondary channel east of the river mouth. This brought it into Ise Bay, where it was shunted eight kilometers further east to the short estuary that led up to the Shiratori yards on the south edge of Nagoya. When the raft was safely anchored in the reception area, the head raftsman was given his receipt, and he and his crew started for home. Whether they took the long way back by boat or headed directly overland to Enjōji is unclear, but one suspects they enjoyed a taste of Nagoya's city life before leaving town.

Figure 14. **Beaching timbers at Shiratori.** At the Owari han lumberyard at Shiratori, men guide a section of raft into position for disassembly and beaching. This section carries cargo under a small shelter. On the far side of the raft is one of the vessels that escorted it down the Kiso. At left laborers using shoulder poles carry a timber up a log ramp for stacking to dry while others return for their next load. At upper right two men stand in front of the guard shack. Reproduced from Tokoro Mitsuo, *The Wood-Cutting and Transporting System of Kiso* (Tokyo: Tokugawa Institute for the History of Forestry, 1977), scroll 2, plate 17. Courtesy of the Tokugawa Institute and its director, Ōishi Shinzaburō.

MARKETING AND SHIPPING TIMBER BY SEA

Members of the timber overseer's staff at Shiratori inspected rafts of merchant logs, charged appropriate user fees, and released them to their owners. For government wood, workmen disassembled the rafts, separated the sticks by type, and measured the volume.[39] They hauled some pieces, probably smaller ones, ashore for storage in the yards. Others they left floating in the holding area. The former were stacked carefully for air drying; the latter, covered with lumber or rocks to hold them under water and prevent their drying.

Government timber was stored until used or sold to lumber merchants by special sale or bid. Special sales provided select high-quality wood for specific purposes at premium prices, as, for example, decorative lumber for use in a teahouse or mansion. Most wood, however, went at prices determined by a type of adjusted bidding.

In this latter method merchants submitted bids to the officials managing sales, and the officials originally used these bids as guides in setting basic fixed prices for their material. Resale prices of wood in merchant lumberyards actually fluctuated sharply in face of erratic shifts in supply and demand, as noted in chapter 2, and this price-fixing mechanism may have helped modulate those fluctuations. However, the longer-term rise in lumber costs that resulted from growing scarcity and inaccessibility of supply meant that prices at Shiratori chronically ran behind those of the resale market, to the disadvantage of Owari's treasury and the lumberyard operation. Consequently, in 1745 *han* officials adopted a policy of annually surveying prices in the yards of lumber merchants. The figures they obtained provided benchmarks against which they adjusted upward their own prices in a clear attempt to shift merchant profits to themselves. Whether they were successful or, as seems more likely, simply added to the inflationary pressure, is unclear.

Government timber at Shiratori went both to and beyond Nagoya. Much moved directly into merchant lumberyards in the city. In such a case the merchant submitted his purchase order, clerks completed the transaction, yardmen measured and marked the wood in the water or storage area, and laborers floated or carted it to the merchant's yard for resale to consumers. In the case of goods bound for other cities, the sale was arranged, the goods measured and marked, and stacked pieces put into the water again. Workmen bound them into rafts, which they poled out to a ship and loaded aboard, although they might lash exceptionally large logs, such as those used in pillars, alongside the vessel. The loaded ship then began its treacherous journey, usually to Edo or Osaka, at the end of which the pieces were disgorged, reformed into rafts, poled to the lumberyard of the merchant, and in due course sold to the consumer.

Recapitulation and Analysis

This elaborate system of timber transport constituted a response to the relentless destruction of timber stands that forced loggers to fell ever more distant and inaccessible forests. The difficulties of transport that accompanied the advance into mountainous terrain were exacerbated by the intensifying competition for land and river use that attended social growth during the seventeenth century.

Lumbermen attempted to overcome these obstacles by developing elaborate administrative systems for managing forests and their harvest. They pursued riparian work to improve the efficiency and reliability of river use, and they applied to their work increasingly complex and extensive mechanisms of transport: sledding, winching, bridging, and chuting and boom, storage, and raft technology.

Figure 15, which brings together figures on labor use mentioned in the text, highlights certain aspects of the transportation system. The column of derived figures (manpower equivalents) must be viewed as nothing more than local and tentative indicators of the relative labor intensity of the several stages of timber transport, but it does highlight the great differences in manpower efficiency of these stages.

By far the most labor-intensive stage of logging was *yamaotoshi,* the tedious process of moving pieces of wood down and out of the jagged and precipitous mountains that formed so much of Japan. And it was precisely this part of the transport system that expanded the most as loggers worked their way deeper into mountain interiors. One result was the development of the chute-trestle technology that moved logs from stump to stream.

Figure 15. Labor use in the successive stages of Kiso timber transport

Process	Men	Koku	Distance (km)	Time (days)	Manpower Equivalents[1]
Sledding	3–6	9	5–6	daily	9
Yamaotoshi	24–40	33–75[2]	150–1,000 (ft.)	daily	329
Shokokugari	64–88	800–1,200	0.5–2.5	daily	5
Ōkawagari	65	20,000–30,000	80	90 days[3]	0.29
Rafting					
Nishikori-Kaneyama	2 (5)	25–30	4	0.25 day	0.45 (1.14)[4]
Kaneyama-Inuyama	1 (4)	25–30	20	0.75 day	0.14 (0.54)
Inuyama-Enjōji	1 (7)	50–60	16	1.0 day	0.11 (0.80)
Enjōji-Shiratori	8 (158)	750–1,800	50	6.0 days	0.08 (1.52)

1. Manpower required to move 100 *koku* 1 kilometer per day. These figures are based on average figures for each column (e.g., for sledding, 4.5 men move 9 *koku* 5.5 km daily, so it takes 9 men to move 100 *koku* 1 km per day).
2. The figures 33–75 assume about 2 to 3 logs per *koku*.
3. The figure of ninety days, which was the season limit, assumes that a crew could launch about 600 small logs (300 *koku* of wood) per day and get the lot to Nishikori without serious interruption.
4. These parenthetic figures are the manpower requirement of rafting when one includes the work force of 110 men and 40 officials at Nishikori, crediting them with processing 50 Nishikori rafts per day (or 25 Inuyama double rafts or 1 full-length Enjōji raft).

Even allowing for losses en route, free floating *(ōkawagari)* was, by contrast, extremely efficient—more efficient in fact than all legs of the rafting operation when the manpower at Nishikori is figured into the labor cost of rafting. Clearly, the general shift from floats to rafts was not promoted by the greater labor efficiency of the latter. Other considerations led to that change. One, certainly, was the loggers' wish to reduce log losses from flooding, stranding, and theft, especially as the market value of timber rose. More compelling, probably, was the desire by villagers and rulers to reduce damage to the river and its riparian structures and the desire by loggers to avoid the lawsuits and legal complications that such damage produced.

These latter considerations of riparian damage and litigation carried enough political weight to compel governments and loggers to accept the increased costs of rafting. This adaptation led, in turn, to a second and strikingly complex technological development, the evolution of log booms such as that at Nishikori.

We noted that this transport technology evolved from pre-seventeenth-century practices: it did not reflect or result in any basic technological breakthroughs or transformations. Rather, it was the massive and dedicated elaboration and application of existing techniques, and the corollary development of institutional arrangements to operate them, that enabled the intricately intertwined agricultural and logging activities of the age to coexist, however uncomfortably, and so to meet the enduring need of Tokugawa cities for the yield of both arable and undomesticated land.

4 | ENTREPRENEURIAL LUMBERING IN YAMAGUNI

SCHOLARS HAVE illuminated many facets of the early modern lumber industry, as citations in chapters 2 and 3 reveal, but few studies attempt to treat in its entirety the process of entrepreneurial lumbering that moved wood from stump to street in any particular locality. That overall process is unusually visible in scholarly reconstructions of the Yamaguni timber industry.[1]

Yamaguni, meaning here the watershed of the Ōi River upstream from Shūzan, was a major source of Kyoto city lumber from the ninth to nineteenth centuries.[2] The area is as mountainous as its name implies, with about 90 percent of it still covered today by the forests that have produced high-quality *sugi* and *hinoki* for a millennium.[3] From the heart of Yamaguni straight to the imperial palace (Gosho) in Kyoto is only about twenty kilometers, but getting timber to the city was difficult because the area's topography forced the wood to follow a tortuous river route to market.

In Yamaguni, as elsewhere, the process that funneled wood from scattered logging sites down to consolidated storage points and then outward again to consumers had an hourglass shape. Its basic stages were these:

1. locating timber and arranging for its harvesting
2. felling the trees and reducing the trunks to appropriate lengths and forms
3. transporting the pieces to storage sites—the neck of the hourglass

4. storing the pieces for curing and sale
5. distributing the pieces from wholesaler to retailer, building contractor, or consumer
6. disbursing the income from the undertaking

Most of these stages were labor intensive and required specialized know-how. As no one person was well placed to oversee

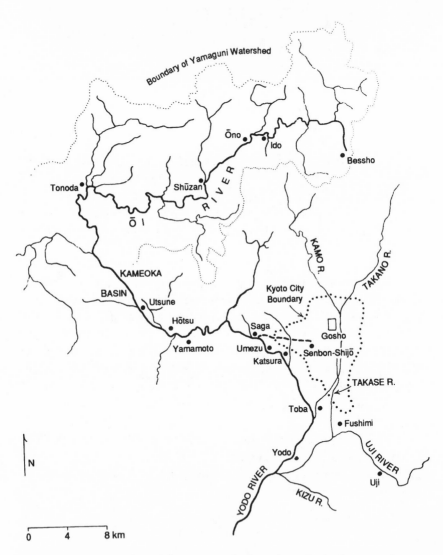

Map 8. **Yamaguni**. Adapted from *Kyōto kubun chizuchō* (Tokyo: Nitchi Shuppan, 1992), map no. 2.

and control, much less to execute, the whole operation, lumbering required the cooperation of several participants. The financing of this multistage, multiparty process proved particularly difficult because, as indicated in chapter 2, it meant that several parties had to commit resources in advance to an enterprise that was inherently risky and could easily fail.

Yamaguni Timber's Route to Market

Timber from Yamaguni normally went to market via the Ōi River—also called the Hōtsu from near Tonoda to Saga, and the Katsura from there to the Yodo confluence. The river follows a long, tortuous course of about eighty kilometers from a source near Bessho village down to the three landings of Saga, Umezu, and Katsura just west of Kyoto. Initially the rivulet flows due north until it encounters a south-flowing source brook. Then it swings west-southwestward, working its way down the central valley of Yamaguni, a convoluted, beautiful valley bounded by abrupt, sharply incised hillsides and leveled at its many bends and confluences by small floodplains of rich, deep, productive soil. On its way through Yamaguni the Ōi receives water from several small branches, but above Tonoda it remains an unprepossessing stream, incapable during dry times of carrying full-sized rafts. Near Tonoda it cuts back southeastward, adding water volume from major tributaries as it flows through the broad, flat-bottomed Kameoka basin to Hōtsu. Below Hōtsu it plunges back into the kind of hilly terrain from whence it emerged at Tonoda, cutting its way through several kilometers of twisting, ledge-studded, white-water gorges before opening out into the Kyoto basin at Saga. From Saga it meanders past Umezu and Katsura, then flows sedately southward into the Yodo River and on to Osaka and the sea.

A lot of wood went from Yamaguni to Kyoto via the Ōi. By one estimate the river carried out some 130 rafts of Yamaguni timber per year in the sixteenth century, the number rising to 200–300 by the 1670s. Output may have stagnated or even declined for some decades after that, but plantation silviculture, mostly of *kabusugi* ("stump cryptomeria," which produced pole timber as coppice growth), increased Yamaguni timber production during the eighteenth century.[4] The number of rafts rose to about 500–

600 by circa 1750–1770, and to roughly 800–1,000 from the 1790s until the late nineteenth century.[5]

This river traffic was not spread evenly through the year; most rafts went in autumn or spring, few in midwinter when stream flow tended to be low and river work most chilling.[6] Throughout the nineteenth century, workmen in Yamaguni were launching five to ten rafts daily during the most intensive periods of spring and autumn rafting. At any given moment in peak season, therefore, as many as twenty-five to fifty Yamaguni rafts (plus a substantial number from other sources) might be bobbing along the river bound for the three landings.[7]

The Ōi proved to be a serviceable lumber artery, but it was far from ideal.[8] As noted in chapter 3, snags, sandbars, ledges, and rock pileups severely limited its carrying capacity until the dredging-clearing work of Suminokura Ryōi and local lumbermen made rafting possible up to about Shūzan. Above Shūzan elaborate wooden dams *(iseki)*, which villagers maintained to divert water to rice paddies, permanently bisected the stream at intervals of a mile or two. An overflow gate with a spillway wide enough (two *ken* or twelve feet) to accommodate log rafts surmounted each dam, but when farmers were irrigating, they scheduled the opening and closing of gates according to local water needs and village agreements. Any raftsman wanting to use the river then had to time his movements to coincide with these maneuvers. When the overflow gate was not in use, rafts slipped over the dam and down the spillway (called a raft chute or *ikada suberi*) with the overflow. At times of especially low water, raftsmen sometimes closed a gate, raising water level upstream sufficiently to float stranded wood into the holding pool. Then they knocked the gate open to create a head of water that could flush along *(teppō nagashi)* any rafts aligned behind the gate or stranded below the dam.

On the several branches of the upper Ōi, lumbermen used smaller permanent splash dams to float and flush along loose logs and small rafts. Some of the dams were wooden structures. Others were rock-filled earthen barriers formed over large, gate-controlled, rectangular wooden culverts through which flowed sticks of wood and water for flushing stranded rafts. Whereas rafts on the Ōi were of standard eight-foot width (one *ken*, two *shaku*), those on the branch rivulets were only half-width. Even then they tended to accumulate vegetation ripped from the edges of the slender streams as they surged along, propelled by the periodic rushes of discharged water.[9]

When the short, half-width rafts reached landings on the Ōi, local raftsmen bound them together to achieve regular breadth, and near Tonoda they linked enough sections together to form a full-sized Ōi raft. About 180 feet long and consisting of twelve lengths of timber, such a raft usually numbered some 275 sticks. To estimate conservatively, it contained about a hundred maritime *koku,* or roughly a thousand cubic feet of wood.[10]

Crews of two or three men working in relays maneuvered each raft from its point of origin, advancing from landing to landing, and ideally arriving at Saga five days later. At each successive landing the replacement crew received both the raft and a manifest that listed the types of wood it contained, the number and dimensions of pieces, the shipper, the receiver, and the rafting organization that was handling its movement. The old crew then headed back upstream on foot, and the new one took the raft on the next leg of its journey.[11]

Just above Hōtsu, at the small landing of Utsune, Kameyama han maintained a transit tax *(unjō)* station where it collected a 5 percent tax on each raft. The fee compensated for damage that rafts and loose timbers regularly did to bridges, riverside levees, and irrigation dams and inlets. The raft-operators' organization *(ikada ton'ya nakama)* of Hōtsu and Yamamoto villages provided the inspectors who checked each raft's manifest and the brands on the sticks of timber, thus assuring that the tax was properly assessed and paid. Once the check was complete, a three-man rafting crew from one of the two villages took charge of the raft and moved it down to their own landing. There they tightened it up for the treacherous run ahead, and when water level and other conditions seemed right, they headed out, bouncing, twisting, and slithering their way through the gorges. If all went well, they arrived at the designated landing a few hours later. There they turned the raft and manifest over to the appropriate merchant recipient and gave thought to the scenic trek home.[12]

Timber thus reached the landings at Saga, Umezu, and Katsura, the "three landings" (*sankatsu* or *sankasho,* lit. "three places") as they were called. But that, of course, was hardly the end of its journey. Indeed, that merely brought the wood to the

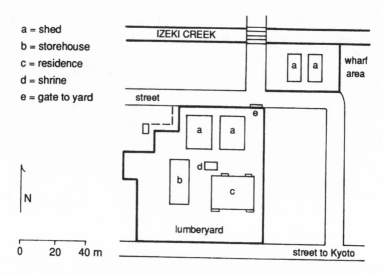

Figure 16. **A wholesale lumberyard at Saga.** The lumberyard of the fifty-two-village lumbermen's guild at Saga, represented here, probably was typical of three-landings lumberyards in its general layout, size, locations of sheds for storage, fireproof warehouse, business residence, and household shrine. Goods were hauled out of Izeki Creek, a lumber canal that carried material eastward from the Hōtsu River. From storage in the yard, the lumber usually continued eastward by cart along the street to Kyoto, but on occasion it was returned to the Hōtsu and sent downriver to Toba for transit to the city via packhorse, cart, or the Uji-Takase waterway, depending, in part no doubt, on the size of the goods and their destination. Derived from Fujita Yoshitani, *Kinsei mokuzai ryūtsūshi no kenkyū* (Tokyo: Ōhara shinseisha, 1973), 390.

narrow point in the hourglass, the yards of the three-landings timber wholesalers (Sankasho zaimoku ton'ya). At the landing the crew beached its raft, and the wholesaler's workmen disassembled it and shunted the pieces down a canal to the appropriate wharf area. They might store the timber in the water pending sale, but more commonly they hauled it ashore and stacked it to dry in the lumberyard, either horizonally in the open or vertically under high-roofed sheds.

The three landings were more or less adequate to their task, but they all had shortcomings. Saga, being farthest from the city, entailed the highest cartage fees for city lumber merchants, and it was not until 1863 that the Saga wholesalers overcame this disadvantage by constructing a slender, fast-flowing canal to carry their goods to Senbon-Shijō near the city center.[13] But, precisely because Saga was farther upstream, it was the preferred destination for the Yamaguni shippers, who wished to minimize their delivery time and cost, and for the Hōtsu raftsmen, who wished the shortest return trip.

More important, Saga was spared a problem that perennially debilitated the other two landings. River silting and shifting, which upstream deforestation and river clearance had exacerbated, forced villagers in the low-lying Umezu-Katsura vicinity to construct and maintain flood-control levees. The river persistently eroded those levees and continually damaged the channels and beaching areas at the two landings, repeatedly forcing wholesalers there to dredge and repair their waterfronts. The problem even led the Umezu wholesalers to abandon their landing for several years early in the nineteenth century. Repair work in the 1820s reopened Umezu, but by the 1840s both landings again were suffering from silting and bank erosion, so the wholesalers again taxed adjoining hamlets, which depended on the lumber work for labor income, and used the receipts to hire their residents to fix the river.[14]

By the eighteenth century these problems of Umezu and Katsura had enabled Saga to become the dominant landing, but wholesalers at all three found their business difficult to pursue. Indeed, damage to the landings and obstacles to overland transfer often led them to reassemble marketed goods into rafts, which they sent on down to Toba village, where members of the Toba carters' guild (*kurumagata*) hauled them up to the city lumberyards.[15]

Some Yamaguni timber went via the three landings to merchants in Fushimi and Osaka, but the wholesalers sold most of it to lumber merchants *(zaimoku ton'ya)* in Kyoto. Merchants situated in northwest Kyoto generally bought from Saga; those to the southeast, from Umezu and Katsura. When the landings were functioning properly, the city merchants hired carters to haul their goods eastward the few kilometers to their own yards. But if a merchant was simply serving as a broker *(zaimoku nakagai),* the goods might go directly to the purchaser, whether a retailer, building contractor, or, in the case of government timber, an official lumberyard.[16]

The process of moving wood from Yamaguni upland to city outlet thus involved collaboration among several groups of people. As the Edo period progressed, however, industrial disputes appear to have become more intense, making that collaboration difficult. Disputatiousness was encouraged by the diffusion of entrepreneurial know-how and the accumulation of capital here and there, which encouraged outsiders to seek a role in the business and participants to try to cut one another out of it. Disputes may also have been fostered by deterioration in lumbering's profitability, although instructive statistics are hard to come by.

To adumbrate the profitability issue, Yamaguni timber supplies became more scarce in the late seventeenth and early eighteenth centuries, driving the cost of lumber production upward.[17] The government, however, tried to keep prices down by legislative fiat. The resulting cost-price squeeze made it imperative that entrepreneurs obtain a reliable estimate of the cost and income prospects of logging projects, but the deterioration of timber stands made accurate forest mensuration particularly difficult, and as a consequence loggers found it difficult to predict costs and yields accurately. One suspects that the heightened cost-price squeeze and difficulty in mensuration fostered the repeated attempts by participants to bypass one another in the provisioning work. At the same time, however, the squeeze forced them to continue depending on one another for the greater amounts of capital that lumbering required.

These entwined patterns of exacerbated rivalry and interdependence are visible in the historical record of Yamaguni lumbering. As Fujita Yoshitami has shown, the lines of socioeconomic tension in the industry were nearly as numerous and interwoven as the number of active and potential participants. The issues that underlay this tension show up most clearly when one looks at the system in terms of its main participant groups, examining them sequentially from forest to city.

The Yamaguni Lumbermen

In the 790s when Emperor Kanmu required timber to build his new imperial capital at Heian, he designated Yamaguni as imperial forest (*gosoma goryōchi* or *kinri goryōchi*).[18] In later centuries the imperial court retained the area as estate land (*shōen*), and those resident holders of woodland who had the customary legal function of purveying wood to the court managed lumbering operations in the dozen villages that comprised the Yamaguni area. Their descendants retained that function, and as of 1600, about a hundred locally prominent households controlled most of the woodland in the villages of Yamaguni. They claimed that their customary function of purveying wood, their "axe right" or "axe duty" *(ono yaku),*[19] as it was known, dated back to the Heian period and had been confirmed by Hideyoshi in his land survey of 1596.

In subsequent decades the number of households involved in the lumbering fluctuated, and axe rights were frequently bought and sold, a process that generally worked to the advantage of larger landholders by increasing both the proportion of Yamaguni woodland under their control and the exclusiveness of that control. By the later seventeenth century most axe-right holders were well-to-do farmers with two to eight hectares (five to twenty acres) of woodland per household, which they harvested as a business supplementary to their tillage work. They also handled logging projects in other Yamaguni woodland, most notably the extensive quasi-communal village lands known as *yakuyama* or "axe-right" land and *miyayama* or shrine land. In toto they dominated both logging and timber-rafting activities in the upper Ōi region.[20]

During the seventeenth century these axe-right holders came to control Yamaguni wood production through guildlike associations (known variously as *kumiai, dantai,* or *nakama*) that negotiated shipping and lumbering contracts with downstream raftsmen and timber merchants. By the 1670s, they had joined their fellows in villages downriver as far as Tonoda to form the fifty-two-village lumbermen's guild of Tanba Province (Tanba gojūnikka mura zaimoku shōnin dantai). Within another half century the guild had secured official recognition as a *kabu nakama,* each member holding a *kabu* that carried designated privileges and obligations and that could be inherited, transferred, or sold in accordance with stipulated procedures.[21] The *kabu* holders used their organization to manage forest labor arrangements, river maintenance work, raft operations within Yamaguni, and the allocation of government fees. The *nakama* also handled relations with outsiders, notably the Hōtsu-Yamamoto raftsmen, three-landings wholesalers, and Kyoto lumber merchants. It excluded other Yamaguni villagers from entrepreneurial roles in felling and rafting work, and it enabled *kabu* holders to discourage internecine price cutting or other mutually disadvantageous dealings.

By the late 1750s, however, the organization had lost its effectiveness and was failing to either control or accommodate other local lumbermen, who were shipping wood without authorization. To cope with this problem, the Tanba lumbermen established a new system of rafting licenses *(ikada han kabu)* unattached to old axe rights. In essence they hoped to restore their guild's control of prices, wages, and market conditions by allowing successful villagers to purchase licenses and join them in denying the river to others.[22] The strategy appears to have proven effective, and Yamaguni lumbermen used it thereafter to protect their common interests vis-à-vis outsiders.

The advantage of the *ikada han kabu* system was that it separated the right to fell and ship wood from the old axe right, thereby offering membership in the organization to local people of other backgrounds. Their inclusion had became necessary because intensifying timber scarcity had given rise to *kabusugi* afforestation, which enabled new households to move into the lumber business at the expense of venerable axe-right holders. This could happen because most of the afforestation occurred on accessible sites near villages and streams, mostly woodlots in the possession of individual households, whereas most of the woodland controlled by axe-right holders (primarily *yakuyama* and *miyayama*) was on interior hillsides with poor access to the river system. As plantation output rose, the wood production of axe-right holders declined in relative importance, and the holders of *kabusugi* stands, regardless of their sixteenth-century ancestry, gradually came to constitute the dominant force in Yamaguni production.[23]

These commercial Yamaguni lumbermen—men like Nogami Rihyōe, mentioned in chapter 2—financed their afforestation and lumbering operations from several sources. Their own logging income, agricultural production, or profits from money lending sometimes sufficed. A logger might obtain funds from a local mutual-aid society *(kō)* to which he belonged. He might borrow from his village or from a moneylender. Or he might receive an advance *(maegashi)* from a three-landings wholesaler or Kyoto lumber merchant. Very commonly, in an arrangement sometimes known as *nenkiyama* (fixed-term land), a Yamaguni lumberman would purchase for eventual felling a young stand planted by a village landholder.[24] He would make a down payment at time of purchase and pay the balance after the lumbering was done, thus having the landholder assume a share of the overall cost in return for an early recovery of his initial afforestation expense.

Whatever the source of his funds, after the lumberman felled his trees, he sent the wood to market via the Ōi. From his viewpoint the river was a mixed blessing. On the positive side, the rafting schedule fitted felling practices nicely. Most of the timber in Yamaguni was *sugi,* and loggers found that *sugi* bark, which they marketed for roofing and other uses, was easiest to peel off in summer and early fall (sixth to midninth months). Moreover, they believed that *sugi* felled then yielded the best quality wood.[25] Loggers could ship the timber as soon as it was felled, barked, and moved to streamside, or they could hold it until early spring, when snow melt and seasonal rain raised the river, facilitating transport. Because the men who rode the rafts often were those who worked at the felling sites, this arrangement enabled villagers to spread their labor time more evenly through the year.[26] As a record of the 1750s put it,

> These villages being in the mountains, forest work is their number one employment. Every year from midsummer to early autumn, villagers fell *sugi* and *hinoki,* producing bark and other roofing to meet village tax requirements while providing themselves with living wages. And from winter to spring, during slack season, they transport goods to Kyoto and other nearby places and sell them to firewood shops, general stores, lumber merchants, bamboo shops, or whatever.[27]

The serendipity of felling and rafting schedules notwithstanding, Ōi rafting was never fully satisfactory to the Yamaguni lumbermen. Besides the *unjō* payments that they made at Utsune from the 1640s onward, after 1676 they had to pay an additional fee on each raft that reached the three landings to defray the cost of periodically cleaning out the gorges below Hōtsu.[28] On top of these regular costs of operation were occasional charges for special repairs. In 1858, for example, the irrigation dam between Ōno and Ido villages had to be replaced. The estimated cost for material and labor was 21.15 *koku* of rice, a sum equal to the annual rice production of even the more wealthy farmers thereabouts.[29] The villagers, who used the dam for irrigation, paid most of that bill. However, the organization of Yamaguni lumbermen was responsible for those parts of the structure necessitated by their raft traffic, notably the raft chute below the dam and probably the gate or at least the reinforcing pieces around it. As most of the lumbermen were also rice farmers, moreover, they had to pay for the dam work twice.

There were other costs, too. At the start of each rafting season, before dispatching the year's burden of timber, the lumbermen had to pay workmen to travel down the river and clear it of debris. In addition, they were continually having to spend sums for litigation against the Hōtsu-Yamamoto raftsmen or the three-landings wholesalers.

As the eighteenth century advanced, the growing number of rafts created additional problems. All rafts had to stop at Utsune for tax assessment and to be taken over by the Hōtsu-Yamamoto raftsmen. During the busy weeks of autumn and spring, rafts accumulated at Utsune, and if not cleared out before the end of rafting season, they had to lie over until the following autumn. Such delays increased the extent of rot and damage, reducing the market value of the timber crop. Whether the bottleneck was caused by a shortage of Hōtsu-Yamamoto raftsmen, by their wish to spread employment more evenly through the rafting season, or by their concern to maintain a safe distance between rafts as they negotiated the gorges is unclear, but to the Yamaguni lumbermen the situation was a cause for complaint. Hoping to solve the problem and cut rafting costs (and perhaps hoping to break Hōtsu-Yamamoto control of the gorges), in about 1764 the Yamaguni lumbermen petitioned to have the tax station moved downstream to Saga. The petition was denied, and instead the problem worsened during the nineteenth century as the number of rafts continued to rise.[30]

The gorges above Saga, where today professional boat operators give summer tourists unthreatening thrills, were a problem in other ways. First, they were frequently the site of raft smashups, with commensurate loss of timber and occasionally life. In lawsuits submitted to the Kyoto city magistrate *(machi bugyō)*, the Yamaguni shippers periodically blamed the smashups on the raftsmen, but they failed to win remuneration or any sharing of the liability.

Second, the Hōtsu-Yamamoto raft guild charged labor fees that reflected the special risks their work entailed.[31] The Yamaguni lumbermen filed a suit to secure reduced fees, but the raft operators were able to justify the costs in a case argued before the magistrate, and they not only retained their wage rates but also thwarted all efforts by Yamaguni raftsmen to run the rapids themselves.[32]

The Yamaguni lumbermen tried to reduce their unit shipping costs by unilateral action as well as legal maneuver. On occasion they dispatched oversized rafts loaded with fuelwood, charcoal, and other marketable produce. In more lawsuits the Hōtsu-Yamamoto raftsmen resisted these measures, partly because the measures made their work more dangerous, partly because they reduced their income per unit of product value rafted. By and large the raftsmen were supported in their resistance by government officials, who doubtless saw these measures as reducing their tax yield while incurring more risks of damage along the river. Although satisfactory figures are not available, the long-run trend in rafting costs seems more likely to have been a modest increase than a decrease.[33]

These several disadvantages in using the Ōi prompted Yamaguni lumbermen to seek alternative routes to Kyoto. Small pieces (board stock and split wood: *itamono, kureki*) could go overland directly from Yamaguni to the city, and this route periodically attracted lumbermen, especially when logging near Bessho. However, the mountain passes are forbiddingly high, as a modern-day bicyclist can attest, so the overland route was of no help in moving large pieces, and throughout the Edo period the Ōi retained its hegemonial role as Yamaguni timber's route to market. *Faute de mieux,* therefore, the Yamaguni lumbermen relied on the Ōi to market their goods in Kyoto, and that reliance forced them to work through both the Hōtsu-Yamamoto raftsmen and the three-landings wholesalers.

The Hōtsu-Yamamoto Raftsmen

The raft operators of Hōtsu village (Hōtsu mura ikada ton'ya) were the larger landholders in the village, and they employed neighbors and kin on their rafts. During the 1590s they acted as official purveyors *(goyōgakari ton'ya)* to Hideyoshi and collected *unjō* fees for his government. They continued their raft work in following decades but increasingly found themselves at odds with upstream lumbermen. In 1672, about the time the lumbermen were forming their fifty-two-village association, the raft operators of Hōtsu organized themselves and raftsmen from Yamamoto village across the river into a *nakama*. The guild's main purpose was to thwart both the maneuvers for lower rafting fees and the attempts to break Hōtsu-Yamamoto control of the gorges. To that end, they handled all negotiations with the village lumbermen above and the three-landings wholesalers below. Guild members adopted a *kabu* system that fixed their number at seventeen, fourteen operating out of the Hōtsu landing, three out of Yamamoto. They also drafted regulations to police their membership's behavior, strengthen their control over fees and work rules, and fight off challenges to their position from other villagers or outsiders. Periodically the organization also had to resolve differences arising between the fourteen members from Hōtsu and the three from Yamamoto.[34]

Nominally the raft operators, like axe-right holders in Yamaguni, represented their villages in facing outsiders, but like them they encountered chronic problems persuading their neighbors that their efforts actually benefited everyone. From their viewpoint the "proof" of their beneficence lay in a triad of facts: they employed fellow villagers in their operations, maintained a fee schedule that seemed to pay good wages while bringing an annual income into village government coffers, and preserved that lucrative leg of the rafting work for members of their villages. But other considerations—for example, that they were generally more well-to-do than their neighbors, that they carefully limited the number of households eligible to participate as entrepreneurs in the trade, that on occasion they agreed to changes in rafting procedure that made the work more dangerous, and that most of the money they fed into village coffers was consumed by litigation defending the raft organization against outsiders—preserved solid grounds for neighborly distrust, much as in the Yamaguni villages.[35] In consequence of recurring internal difficulties, the negotiating strength of the raftsmen's guild waxed and waned. It did survive, however, playing a key role in Ōi river lumbering for the rest of the Edo period.

The Three-Landings Wholesalers

At the next stage of lumber's journey were the three-landings wholesalers.[36] Like Yamaguni axe-right holders and Hōtsu-Yamamoto raft operators, they were essentially the largest landholders and most influential members of their villages. Also like them, they asserted a special right to their jobs, claiming they had provided government lumber to Kyoto from even before the days of Hideyoshi and Ieyasu.

During the seventeenth century they collaborated unofficially as the three-landings timber wholesalers' guild (Sankasho zaimoku ton'ya nakama). Doing so enabled them to deal as a body with the Yamaguni and Hōtsu-Yamamoto groups above and the Kyoto lumber merchants below, as well as with would-be rival wholesalers from nearby villages and occasional timber buyers from Fushimi, Osaka, or elsewhere. They also worked through the guild to keep the gorges open, maintain their landing facilities, organize their local labor supply, and resolve differences between the dozen or more wholesalers at Saga and the dozen or fewer at the other two landings, Umezu and Katsura.

In 1676 representatives of the wholesalers and of the recently formed fifty-two-village lumbermen's association set their seals to the resolution of a dispute over marketing arrangements and the funding of river maintenance. The agreement contained these three clauses:

> At this time we agree that for every *kanme* [one thousand *monme*] required in river work in the gorges, the upper Ōi villages will provide seven hundred *monme* and the three landings, three hundred.

> To decide the dues of each person, a set fee for each raft will be levied and a person's calculated total turned over to the guild. This being agreed, henceforth whenever there is repair work in the gorges, the village's calculations will be made in this manner, and the figures on repair costs will be duly reported by the three landings in an account ledger.

> The two parties in assembly have settled the handling of river expenses in this way. In addition, merchants from the three landings will not engage in buying directly in the upper Ōi region, and lumbermen from there will not engage directly in marketing at the landings. To these commitments do we now set our seals.[37]

Thus, at the end of a document primarily concerned with paying for river maintenance, the two parties struck a deal in which the former agreed to buy timber in the upper Ōi region only through the association while the latter agreed to sell only through members of the wholesalers' organization. Nevertheless, before long the Yamaguni lumbermen were eroding the agreement by again selling goods directly to others, and by century's end they had established branch offices of their own in the Kyoto vicinity. That development, together with stagnation in Yamaguni wood production, forced more than twenty of the original thirty-three *sankasho* wholesalers out of business, creating vacant seats *(kyū kabu)* within their organization and, in all probability, fostering a sense of desperation about their business prospects.

Despite their cooperation as an informal guild, the majority of wholesalers thus were inactive by the early eighteenth century, and the fifty-two-village association had established its own outlets in the Kyoto area. During the 1730s the wholesalers tried to retaliate. In 1732 they moved to trim their losses by unilaterally reducing the prices they proposed to pay for Ōi timber and by halting further purchases. Presumably they were hoping to put pressure on the villagers and stanch the drain on their own resources while creating a downstream scarcity that would increase the market value of the lumber they already had in stock. The outcome of that maneuver is unclear, but it seems to have failed, probably because the Yamaguni lumbermen simply sold to other buyers. In any case, in 1734 the wholesalers tried a different tack. They struck a deal with the Kyoto city magistrate in which they agreed to defray certain expenses of the magistracy in return for official recognition of their *kabu nakama* status and assurance that thenceforth no one else would be allowed to deal in lumber shipped to Kyoto via the Ōi. The transaction was basically similar to that arranged by Edo lumber merchants and their city magistracy at about the same time.

This agreement converted the informal guild into an officially recognized oligopolistic group and emboldened the wholesalers to file a lawsuit enjoining the upstream lumbermen from bypassing them and selling directly to city lumber merchants. The city merchants lobbied on behalf of their clients, however, and the fifty-two-village lumbermen themselves gave as good as they got. In rebutting the wholesalers' suit, they boldly claimed, in cheerful disregard of their 1676 agreement, that as a quid pro quo for the *unjō* payments they had always made at Utsune, they had the right "to sell timber freely to lumbermen of Saga, Umezu, Katsura, and also Kyoto, Fushimi, Osaka, or anywhere else."[38]

The dispute dragged on until 1742, when the wholesalers finally agreed to a compromise. It stipulated that the upstream lumbermen were no longer to sell lumber rafts except at the three landings, but they were authorized to take over one vacant wholesaler's seat at Saga and establish a new one at Katsura. Moreover, they could deliver up to half their production to those two outlets; the rest was to go to the regular wholesalers at the landings.

The outcome of this compromise was hardly what the wholesalers hoped. Within a few years the fifty-two-village association had its two lumberyards in operation, handling their share of production. But during the 1750s, even as the two flourished, several regular *sankasho* wholesalers went out of business, leaving their *kabu* inactive or even selling them to pay off debts.

A record of 1758 suggests how unsuccessful were wholesalers' attempts to meet the new competition. At the time some of them were making advance payments to upriver lumbermen as a way to steer trade to their own yards. To do so, however, they were having to borrow the funds from other lumber merchants. In the winter of 1758 one Katsura wholesaler found himself being pressed to repay monies he owed. As required by *nakama* regulations, he reported the circumstances to his colleagues, indicating that he had borrowed 296 *monme* and made advances of 59.2 *monme*. In hopes of retaining his good standing in the guild, he assured members that he would repay the entire principal after marketing his raftloads of wood the following spring.[39] Whether he succeeded is unclear, but if not, he was not alone, as the increase in vacant *kabu* suggests.

It thus appears that nearly a century of sporadic maneuvering had failed to secure the interests of the three-landings wholesalers. Nor did they seem to benefit from the late-eighteenth-century expansion of Ōi river timber traffic. During most of the eighteenth and nineteenth centuries, only two of the six authorized Umezu wholesalers were actually engaged in lumber business at any given moment—and only four or five of the original eleven at Katsura and about six to nine of the sixteen at Saga.[40] Clearly, the great increase in raft volume after about 1750 was not benefiting the three-landings wholesalers as a group. One suspects that those who did manage to stay in business did so by expanding their gross turnover sufficiently to offset reduced profit margins. As the dispute of 1734–1742 suggests, moreover, one reason for their persistent difficulty was that they were being outflanked by Yamaguni lumbermen working in collusion with Kyoto city merchants, the last major group in this provisioning chain.

The Kyoto Lumber Merchants

The Kyoto city lumber merchants were a hybrid group. Some functioned primarily as brokers, arranging sales of timber for the three-landings wholesalers in the manner of timber *nakagai* in Edo and Osaka. Others combined the functions of wholesaler and broker, purchasing, storing, and selling their own timber while brokering that of others. They were scattered through fifteen wards *(chō)* of the city. About half of them clustered in the Senbon-Horikawa vicinity, mostly along the main road from Saga to the imperial palace, which is still known as Marutamachidōri (logtown street), and along Sawaragimachidōri (cypress-wood street) to its immediate north and Takeyamachidōri (bamboo-sellers street) to its south. Other lumbermen lined the Takase River, the north-south canal that could receive wood sent down the Kamo and Takano Rivers, and up which men hauled small, shallow-draft cargo boats, the celebrated *Takasebune,* from the Uji River at Fushimi. The particular sites of purveyors changed as fortunes waxed and waned and fires reshaped the city, but the overall distribution corresponded roughly to the division between those who dealt primarily with the three-landings wholesalers and those who obtained most of their lumber via the Takase from Osaka, the Lake Biwa–Seta–Uji River raft route, and other sources in the Kinai region.[41]

Whereas the three-landings wholesalers were generalists who handled whatever sorts of wood came down the Ōi, the Kyoto wood merchants specialized. The core group was the *motozaimokuya,* who handled heavy timber. In addition there were *hikiitaya,* merchants who apparently dealt in sawn goods such as long boards; *shirakiya,* who mostly dealt in smaller split pieces and semiprocessed wood such as cooperage and shingling (for which *sawara* was much used); *takeya,* who marketed bamboo; and *kobokuya,* who apparently recirculated the wood salvage continually generated by urban repair and reconstruction.[42]

Tension between the three-landings wholesalers and city lumbermen was inevitable, no doubt, given the capacity and readiness of some of the latter to function as their own purchasing and storage agents. Moreover, whereas the wholesalers surely wished to be sole providers to "their" city lumbermen, doubtless the latter wished to purchase from the most favorable sources, whether calculated in terms of price, quality, quantity, or reliability of supply.[43] As timber grew scarce in the late seventeenth century and the cost-price squeeze became more severe, upstream groups formed guilds to protect their customary functions while city merchants maneuvered to bypass them. Dealers in small goods, such as the *shirakiya,* tried to buy and transport overland from

Figure 17. **A cooper at work.** Coopers, such as this one at Fujimihara in Owari Province, often preferred *sugi* wood from Yoshino, but they also used *sawara* and other conifer wood, which they hooped with split bamboo. Loggers met the overall growth in demand for cooperage by exploiting remote areas from which smaller-sized pieces could be removed even when heavy timber could not. From an unidentified partial edition of the series of prints "Thirty-six Views of Fuji," by Hokusai Katsushika (1760–1849), in the author's possession.

Yamaguni. Buyers of heavy goods attempted to purchase raftloads directly on the upper Ōi.

When buyers for these city merchants went into Yamaguni, they, like the wholesalers, offered *maegashi* to individual woodland holders, loggers, and raft operators as a way to attract sellers and commit them to a contract. They could do so, even though they charged interest on the advance, because the recipients of the advance welcomed—or needed—this means of recouping their investment quickly. However, the advancing of funds shifted more of the investment burden from producers to downstream merchants, exacerbating their financial exposure and making them more vulnerable to market fluctuations. By the eighteenth century more and more Kyoto timber merchants were finding themselves periodically strapped for resources, falling into debt, and unable to complete payment on lumber contracts or other obligations.

The problem expressed itself indirectly in 1723. Before that year the *motozaimokuya* had, in return for government recognition of their business privileges, maintained a bridge used in the Gion festival (at Shijō on the Kamo River) and the Kanse Ichidai Nō theatrical stage (which had enjoyed Tokugawa favor since Ieyasu's day). Perhaps because of their worsening economic situation, that year they persuaded the city magistrate to order the *hikiitaya, shirakiya, takeya,* and *kobokuya* to assist in bridge repair. When these groups steadfastly resisted the new imposts, a frustrated magistracy in 1729 purchased the necessary timber and sent the bills to the city lumber merchants with instructions to reimburse the government. The merchants protested, complaining that they were already being driven out of business by competitors who were exempt from such burdens. The three-landings wholesalers presumably were among the alleged miscreants, and the magistracy launched an inquiry.[44]

By then several city lumber merchants were in arrears to the wholesalers while others were having difficulty recovering loans, and they may have been using this issue for leverage in resolving those matters. About 1730, however, much of Japan was convulsed by a period of extreme price gyrations, crop failure, and famine, and the government inquiry seems to have made little headway until early 1734, after conditions had improved. Then the magistrate, still trying to solve his bridge problem, instructed the three-landings wholesalers to report on their public service.

The city merchants hoped this investigation would lead the magistracy to require wholesalers' support of the Gion festivities, but the wholesalers, as noted above, evidently saw in it an opportunity to strike a deal that would restore their position on the Ōi River. After enumerating all the public services they customarily performed, they magnanimously volunteered to help defray costs of the bridge in return for government support of their exclusive function on the Ōi.

The shogunate had previously abandoned its early preference for a pluralistic marketplace, as noted in chapter 2, and its Kyoto agent, the city magistrate, accepted the proposal. City merchants and upper Ōi lumbermen joined forces to protest the wholesalers' maneuver, as mentioned above, and the fifty-two-village lumbermen emerged as the chief beneficiaries of the resulting compromise of 1742. In following decades they flourished while the city merchants continued to have difficulty meeting their obligations to the wholesalers.[45]

Participants in the Yamaguni trade enjoyed a respite after 1788, when the Tenmei fire created a huge demand for timber by ravaging Kyoto. The conflagration consumed the imperial palace, 60 mansions of the highborn, 1,150 temples and shrines, 8,100 warehouses, and, reportedly, 183,000 ordinary dwellings—in short, most of the structures in a city of some four hundred thousand people.[46] The building boom faded by about 1800, however, and the problem of derelict lumber merchants reappeared, worse than ever. *Sankasho* wholesalers' regulations of 1806 and 1814 expressly instructed members to report at *nakama* meetings all cases of arrears by Kyoto lumber merchants so that the names of malefactors could be added to the existing list of sixty to eighty lumber merchants with whom further dealing was forbidden until all their debts were made good.

In following decades financial problems continued to trouble the city merchants, prompting the three-landings *nakama* to warn again in 1830 that members must not sell timber on credit because pervasive arrears were making it difficult for the wholesalers to meet their own obligations to the Yamaguni lumbermen. The government's abortive abolition of the *kabu nakama* system nationwide during the 1840s led to some disruption of Kyoto lumber marketing and a sorting out of relationships. By 1850 only nine Kyoto lumbermen appeared on the wholesalers' black-

list.[47] Doubtless the disorder of the 1860s, which led to more fires
and rebuilding in Kyoto, benefited the city lumber merchants,
along with such others as the already-mentioned Nogami Rihyōe.
However, it was the sort of transient, conditional prosperity that
seems to have characterized Yamaguni lumbering throughout the
Edo period: moments of efflorescence in a life of hustling and
uncertainty.

Recapitulation and Analysis

Entrepreneurial lumbering in Yamaguni involved the cooperation of several groups, most notably the Yamaguni lumbermen
themselves, raftsmen along the Ōi River, timber wholesalers at the
three landings, and lumber merchants in Kyoto. Together these
participant groups provided all the requisite technical knowledge
of silviculture, logging, rafting, wood processing, funding, and
marketing. They were also able to mobilize enough labor and capital to accomplish their complex, multistage task.

Laborers were mostly neighbors and kin of the lumbermen,
and while the work provided a livelihood for some, it appears to
have been a source of supplemental income for most. Regarding
capital, the material presented in this chapter does not show with
satisfying precision just how funding was accomplished. It does
show, however, that problems in capitalization were chronic and
that entrepreneurial risk was spread widely among the participants, from landholder and village lumberman to raftsman,
wholesaler, and broker. It is clear that the economic fortunes of
participants were erratic, that indebtedness was a persistent and
probably escalating problem, and that participants pursued other
enterprises concurrently, usually agriculture and money lending,
thereby reducing their dependence on lumbering. It is surely significant that these economic issues lay at the heart of most disputes among participants.

In the risky world of Yamaguni lumbering, a device regularly
employed to protect participant interests was the claim of a spe-
cially privileged role, one sanctioned by custom and authority.
Most notably, practically everyone in the trade claimed, with
greater or lesser validity (and with a conviction surely sustained
by the argument's convenience), that their functions in the purveying chain dated from earlier epochs and had been certified by

Hideyoshi and/or the early Tokugawa rulers as a quid pro quo for providing *goyōki* or making *unjō* payments.

As the seventeenth century advanced and government provisioning became less central to the industry, lumbermen became more vulnerable to the vicissitudes of the marketplace. Along the Ōi, as elsewhere, that situation prompted the several groups to strengthen their internal cohesiveness and their control over their own segments of the business. As timber became more scarce and dear, maneuvering within and among the groups became more intense, most frequently as groups tried to bypass one another in the provisioning chain. Moreover, various outsiders—such as people not resident in the fifty-two villages of the upper Ōi, villagers near the three landings, and merchants from Fushimi and Osaka—tried to get into the trade as irregular scarcities periodically spiked consumer prices.

By the late seventeenth century the *bakufu* was acquiescing in de facto oligopolistic control of many areas of commerce, and during the 1720s and 1730s it formally recognized such arrangements in the Yamaguni trade. Unsurprisingly, *bakufu* certification of privilege did not end the conflicts. Instead, it seemed mainly to entangle the authorities yet further by prompting disputants to engage in more suits and countersuits. This choice of weapon doubtless helped perpetuate the balance among the groups because it expanded the *bakufu*'s role as mediator, and Tokugawa dispute-resolution policy usually consisted of pragmatic, case-by-case attempts at compromise and reconciliation based on a combination of precedent and attention to the status and needs of the disputants.

The very persistence of the maneuvering from the seventeenth to nineteenth centuries indicates that no single group succeeded in imposing its will on the Yamaguni-to-Kyoto system as a whole. Rather, the struggle for advantage in the context of chronic economic uncertainty led all groups to develop more elaborate organizations that promoted their members' interests. In consequence one long-term trend in the history of Yamaguni lumbering was growing elaborateness of industrial organization.

A related trend seems to have been a growing sophistication of economic and political maneuver, particularly by the Yamaguni lumbermen themselves. That growing sophistication, together with the increased scarcity of timber after 1700, which tilted the basic supply-demand balance in their favor, enabled them to strengthen their hand in the industry and extend their activities to the very outskirts of the city.

These developments were gradual, however, and did not completely dislodge any of the participant groups from sanctioned roles in the overall process. In broadest terms, therefore, it may be best to characterize the diachronic pattern of Yamaguni lumbering as dynamic equilibrium. The dynamism was rooted not so much in a wild entrepreneurial ambition to strike it rich, over the dead body of one's neighbor-competitor if necessary, as in the pedestrian desire to arrange affairs to avoid economic disaster and consequent personal or household catastrophe. It manifested itself in sharply conflicting pressure for and against change in rights of participation in the trade. This pressure was continually exerted, each initiative with enough effectiveness to provoke a response, thereby sustaining tension in the industry throughout the Edo period.

The equilibrium derived from the inability of any one group to vanquish rivals in the struggle for advantage. Instead, they all survived, those jostling for position in the 1860s being social descendants of those who had done so in the early seventeenth century. Their strength relative to one another had changed; their basic positions and functions had not. The long-term change in participation was not of groups but of members of the groups: individuals flourished and failed, and as they did, households rose and fell. But as this process occurred, new households essentially stepped into the commercial roles of those whom they had displaced. The roles gradually modified in various ways; they never disappeared. Perhaps one could call it a pattern of internal dynamism and external stability. Some might see in it similarities to the late-twentieth-century Japanese industrial economy. In any case, it was a set of arrangements that successfully transmuted Yamaguni trees into Kyoto buildings for more than two centuries.

5 | LAST REFLECTIONS

THIS INQUIRY has sought, in the first instance, to illuminate the hunter-gatherer process whereby the builders of early modern Japan's wooden cities, towns, and monuments obtained the timber they needed despite the difficulties that task entailed. It has also tried to place this provisioning activity in the broader context of the era's commercial history as viewed from an ecological perspective.

It may be useful to address briefly the question of how this ecological (or more precisely, homocentric autecological) perspective helps illuminate Tokugawa commercial history.[1] Brevity invites excessive reductionism, but in essence this perspective expands the number of variables and the array of interest-group relationships that the historian must consider. It encourages us to be aware of the broader material context of life and the wider "natural" community of which we are a part. In doing so it alerts us to the potential presence of "losers" even in circumstances where all the humans seem to be "winners," including circumstances in which, depending on their scale and character, the biosystem's losses may foreshadow later troubles for the human component of the community—the dead canary in the mine shaft, so to say.

Mainly this ecological perspective calls attention to the effect on human affairs of changes in the environment, whether those changes result from human action or other factors. An obvious example in this study is Yamaguni, where deforestation of interior

regions altered timber production costs enough to foster plantation forestry and as a by-product to change the social composition
of the Yamaguni lumber industry. More generally, as in the case of
Kiso, deforestation led to the elaboration of provisioning techniques and administrative arrangements. It also contributed to the
petering out of urban construction and the reduction of government involvement in lumber provisioning, with countrywide
effects on the organizing of that industry.[2] That intrasocietal factors mediated and helped shape particular outcomes needs no
belaboring.

In terms of interest-group relationships, advantages enjoyed
by the human-centered community of domesticated plants and
animals—what I labeled the "homocentric biological community"
in the Introduction—generally were gained in Tokugawa Japan, as
elsewhere, at the expense of undomesticated flora and fauna.
Because "wild" goods remained essential to the Tokugawa populace, however, that situation promoted conflict between those
engaged in agricultural and those engaged in hunter-gatherer
activities, which observation applies particularly to lumber transport, as preceding chapters showed. Similarly, although not a subject treated in this study, the spread of plantation forestry shifted
parcels of woodland from natural growth to domesticated timber
crops, mainly *sugi,* and that trend generated conflict between
those seeking fuelwood and fodder and those wanting timber.[3]

When Tokugawa society encountered unprecedented difficulty in expanding its ecological base during the eighteenth century, interest groups within it maneuvered to secure their own
well-being. They competed for control of resources, shifted arable
acreage from one crop to another, restricted consumption by some
sectors of society to assure sufficiency for others, devised socioeconomic arrangements to limit the size of the human population,
and, it appears, minimized society's stock of pets or draft animals
to provide more goods for people.[4]

This ecological approach echoes a main theme of Marxian
analysis in its basic diachronic image of growth giving way to stasis. However, whereas that analysis explains the trend in terms of
class struggle and shifting power relations within society, this
approach situates its explanation in changing environmental conditions that stemmed directly from the overall increase in numbers
of people and collaborating flora during the seventeenth century.

Also, in place of the Marxian concept of feudal-bourgeois conflict that sooner or later culminates in the latter's triumph, this approach accommodates evidence of a complex and prickly interdependence of governing and commercial elites that proved durable—dare one suggest that it has lasted to our own day and is, in fact, normative to industrial society?—despite changes in the character and composition of those social categories.

Rather similarly, instead of the modernizationist perception of a more-or-less continual and societywide rise in standard of living thanks to entrepreneurial dynamism, this growth-stasis formulation envisages seventeenth-century gains that gave way to a century of hardship, heightened regional disparities, and exacerbated social inequalities before yielding to a period of renewed social well-being that commenced about 1800. That renewal recouped some of what was lost in preceding decades, but it did not alter the ecological fundamentals that had produced the earlier stasis, and hence it was severely limited in its potential.

Japan could escape its restrictive early modern environmental context only by achieving a radical break with the past: adopting major new technologies to maximize resource exploitation and massively expanding its resource base by redefining it, in the manner of other industrial societies, to encompass the entire planet. In the century and a half since 1850 those changes have occurred; but for Japan, as for industrialized humankind generally, these measures may simply have projected the time of renewed limits a few decades into the future, and in exchange for those few decades of grace the measures may have profoundly complicated the prospects of a sustainable homocentric biological order, even one as harsh as that of eighteenth-century Japan, ever being achieved again.

Turning to less portentous matters, the Tokugawa lumber industry reflected the broader commercial experience in several ways, most notably in terms of government-entrepreneur relations, the nature of business organization, and changes therein. In some specifics, of course, its history was as idiosyncratic as that of other major fields of commerce. Because timber was so crucial to the construction activities of rulers, the early Tokugawa industry was tied especially closely to government. Later, because government interest in construction diminished so sharply, the shift from command to entrepreneurial arrangements was more pronounced than in such notable commercial arenas as foreign trade and rice marketing, in which the government role remained stronger, or cloth marketing, where it was never predominant.

Another area of idiosyncrasy was the timing of lumbering's evolution. Most of the industries based on arable production, notably the rice trade, continued to expand throughout the seventeenth century, not facing the stresses of sharply diminished growth until the eighteenth. Many crafts seem to have shown an even longer-term growth in activity, while a few major industries, notably cotton, domestic raw silk production, and fisheries, experienced most of their growth during the later Edo period. By contrast, the lumber industry passed from growth to contraction in the mid-1600s. Only gold and silver mining, stone quarrying, and armaments manufacture contracted sooner, the first two for lack of supply, the latter two for want of a market.

The early shift in the trajectory of lumbering reflected the rhythms of both supply and demand. Most monument building and the initial surge of urban construction had ended by the 1640s, by which time many old-growth stands had already been consumed. Government withdrawal from construction work deprived lumbermen of a paymaster and made harvesting of the surviving, relatively inaccessible stands much more risky. So entrepreneurs developed forms of industrial organization and financing that enabled them to exploit the continuing but irregular market provided by urban conflagrations and ordinary decay.

When lumber output expanded again during the nineteenth century, it did so mainly through the rise of plantation forestry at sites closer to urban marketplaces. In this study, the *kabusugi* production of Yamaguni illustrates that trend most clearly, but plantations were found in many other places, most notably the Ōme and Nishikawa timber districts west of Edo, the lower Tenryū River valley, and Yoshino in southern Yamato. In those developments, one sees Japan's lumber industry beginning its transformation from an exploitative hunter-gatherer operation into the regenerative agricultural pattern of modern silviculture. Today that pattern blankets much of the archipelago with aesthetically delightful but economically problematic and ecologically worrisome, even-aged, monoculture stands of *sugi* and *hinoki*.

The oligopolistic, multiparty, multistage provisioning techniques that lumbermen devised to replace government supervision during the seventeenth century organized and stabilized the marketplace sufficiently to make entrepreneurial lumbering possible. During the later Edo period, rural entrepreneurs proved particularly adept at exploiting those arrangements, as well as the new plantation silviculture, to gain considerable prominence in the industry. Their success requires more comment.

An issue that has loomed large in Japanese scholarship on the early modern lumber industry is its meaning in terms of rural-urban power relations. Studies written in the 1950s and 1960s described the Edo period as a time when feudal (command) lumbering was supplanted by capitalist (entrepreneurial) lumbering and when, as a corollary, urban merchants gradually extended their control over the countryside. This view, for example, interpreted the practice of advance payment in logging contracts as evidence that city lumbermen were gaining control over village holders of woodland: villagers, it was inferred, needed money so badly that they were willing to cede effective control of their land to get cash.[5]

More recent research, however, suggests that the change in balance of power most commonly went the other way, with rural elements gradually extending their control over the urban end of the business. In this interpretation, urban timber merchants offered advance payments only because they would otherwise lose their timber supply. They provided such payments, moreover, even though doing so placed more of the investment risk on themselves and further contributed to rural strength and urban vulnerability.[6]

This latter view, which is most evident in chapter 4 of this volume, is grounded in more substantial scholarly research than is the former. It conforms more closely to the pattern scholars have found in other areas of commerce, as noted in chapter 1, and it accords with certain basic truisms of early modern lumbering. Specifically, in the early 1600s urban lumber merchants enjoyed three key advantages vis-à-vis their rural counterparts, advantages they lost by century's end. First, because urban merchants had connections to the rulers through their work as agents in command lumbering, they gained favorable access to government-controlled woodland and building projects. Second, they knew how to handle the urban commercial market as well as official construction work and were therefore well placed to dispose of the timber they acquired. Finally, because timber was plentiful, they enjoyed a favorable market position—the classic buyer's advantage—and could combine that advantage with the others to dictate the terms of provisioning.

By the late seventeenth century, however, the growing scarcity of timber put the rural provider in a stronger position. Concurrently, urban lumbermen lost their other two advantages because government demand no longer dominated the market and because the spread of education into the hinterland meant that rural entrepreneurs were gradually closing the gap in know-how. They, too, learned how to organize themselves, obtain government legitimation, launch lawsuits, and negotiate in the marketplace. By the 1750s, rural lumbermen were infiltrating and manipulating urban lumber markets to their own advantage. Nogami of Yamaguni and Noguchi of the Shinanoya can be viewed not only as examples of successful lumbermen, as they were in chapter 2, but also as examples of how rural lumbermen turned their changing situation to advantage. In the process they expanded the scale and range of their operations and ended up urban lumbermen themselves.

As that last sentence suggests, however, an irony of this analysis is that the rural-urban dichotomy seems to dissolve under close scrutiny. Even Kinokuniya, the quintessential urban lumber magnate, began as a rural entrepreneur. Most Edo-period Japanese were born in the hinterland, so it is unsurprising that most successful families had rural origins, but because the arenas of large-scale economic activity were urban, the large-scale commercial success stories generally unfolded there. By the nature of the timber business in particular, as rural lumbermen flourished, they moved to the cities, which provided the major markets they sought. It should be noted, however, that early lumber magnates such as Kinokuniya often made their initial fortunes in other fields, becoming influential lumbermen only after they had established their urban presence and connections. Later ones, by contrast, used rural advantages to become successful lumbermen, entering the urban marketplace later in their careers.

Because lumbering required the cooperation of people all along the route from forest to city, it always cut across rural-urban distinctions. Rather than juxtaposing town against country, therefore, it may be more useful in thinking about long-term developments to adopt Shimada Kinzō's focus on the lumber industry's expanding scale of operation.[7] In the early Edo period, he writes, an ideal/typical entrepreneur shipped to market timber that he received from independent local producers or that was produced locally from government forests at government behest. Later the entrepreneur became a *motojime,* a sort of promoter-manager, managing his own logging activity while also handling that of other villagers, paying on their behalf any taxes or other required fees. From the mid-Edo period on, such an entrepreneur shipped to the city not only lumber extracted under his own management, but also large quantities of wood that he purchased from independent producers. This meant, says Shimada, that he had assumed the functions both of a timber producer and of a broker or financing agent.

Shimada uses the case of Noguchi of the Shinanoya to illustrate this last stage of the evolution, but Nogami of Yamaguni also would do. And both Yamagataya Kitauemon in southern Shinano in the 1730s and the Yamatoya who handled logging for Numata han in the 1680s illustrate the earlier *motojime* patterns. Such people did commonly begin in the hinterland and end up in the city or town, but for Shimada it is the changing scale of operation rather than the change of residence that is significant.

A second problem with the dyadic rural-urban formulation—a problem inherent in all dyadic analysis—is that it conceals diversity and difficulties within categories and overlap between them. The variety and tensions among urban lumbermen have been very evident in the present study. The rural analogue is suggested in the paragraphs of chapter 4 that treat Yamaguni lumbermen and Ōi River raftsmen. Studies of other areas have also pointed up the ways in which the rise of rural lumbering and afforestation was commonly accompanied by enhanced disparities of local wealth and power. In areas such as the upper Kinugawa and Naguri vicinities of the Kantō and the upper reaches of the Ōi north of Kyoto, these developments appear to have produced a few wealthy local lumbermen who controlled the vicinity's forest economy. In other areas, however, such as the upper Yoshino in southern Yamato, a substantially larger proportion of the local populace continued to share in the control of woodland production.[8]

In fine, one noteworthy trend in the early modern timber industry was the shift from predominantly command to predominantly entrepreneurial lumbering. The shift was never absolute, however, because even though successful lumbermen clearly were entrepreneurs who exploited market opportunities, they continued throughout the Edo period to nurture government connections. They did so because the rulers remained prime consumers who could bestow special privileges and advantages on those they favored.

Another noteworthy trend was the apparent shift of initiative and managerial effectiveness from urban toward rural entrepreneurs. This trend reflected the seller's market created by growing timber scarcity, the declining role of government in lumbering, and the spread of entrepreneurial know-how into the hinterland. As Shimada's analysis emphasizes, this trend was accompanied by a structural evolution from simple to more complex organization and from smaller to larger enterprises. In many localities it was also accompanied by growing disparities of wealth and opportunity. These changes emerged as participants in the trade struggled with the implications of lumbering's inherent irregularity and its enduring need for substantial investment capital. Individual households waxed and waned, and they commonly sustained themselves only by combining lumbering with other enterprises. As a whole, however, these early modern lumbermen met the essential timber needs of their cities and towns for more than two centuries.

To facilitate their provisioning work, entrepreneurs devised not only advantageous institutional and financial arrangements but also technological refinements that enabled them to bring timber out of deep mountains and down tortuous rivers to urban markets. The elaborate winch-chute-raft-storage technology of chapter 3 was central to their business. They did not, however, employ certain devices that might have permitted them further efficiencies of production. Three deserve special comment: lumber wagons, crosscut saws, and water-powered sawmills.

Packhorses and drivers were plentiful in Tokugawa Japan, but the ubiquity of steep mountainsides sharply reduced the utility of both carts and wagons in woodland itself, and the official prohibition of wheeled vehicles on highways precluded cartage there.[9] Carts, usually pulled by humans, were permitted in cities; but the heavily laden carts that hauled lumber, stone, and grain about town chronically broke through bridges, rutted wet streets, damaged fences and walls, and obstructed the passage of others. These problems led to city regulations restricting the weight, size, and deployment of carts; persistent violation of the regulations; and repeated quarrels and lawsuits.[10] Wagons, which can bear much heavier loads than carts because they have two axles and can use teams of horses or oxen, appear never to have been used. Prohibited on highways, they would have been nearly useless in town, once the urban layout was established, because of the narrowness and sharp corners of most city streets. Initially created by political figures for political purposes, that is to say, Tokugawa highways and cities proved inhospitable terrain for lumber wagons and even for cartage.

Regarding crosscut saws, by the late fifteenth century blacksmiths had developed large saws with sufficient temper and consistency of thickness to function as crosscut saws, and woodsmen used them to fell timber in Hideyoshi's day. In following decades, however, governments outlawed their use, evidently to discourage illegal felling: whereas saws cut quietly, the sound of axe blows can alert a forest warden or village headman to mischief afoot.[11] Later still, as old-growth forest disappeared, more and more logging involved small trees that were felled easily enough by axe—even the slender-hafted, short-bladed axes favored by Tokugawa woodsmen—to reduce the appeal of crosscut saws.

The matter of water-powered sawmills is less straightforward. Waterwheel technology was long known in Japan, and during the latter half of the Edo period, power generated by waterwheels came to be used in many places to mill grain, press oil seeds, and pulverize material for ceramic work.[12] Frequently, however, these operations became objects of civil lawsuits because they disrupted stream flow, fouled water supplies, and damaged downstream irrigation systems. In consequence, in some places the construction of waterwheels was prohibited and in others the timing of their use was sharply restricted to protect downstream interests.[13] This general situation did little to encourage the use of water-powered sawmills.

Moreover, water-powered sawmills are not simple devices to build or operate. To the contrary, they are massive but elegant contraptions that can involve intricate combinations of wheels, gears, drive belts, and levers to operate the saw blade (whether circular or vertical-cut) and to grip and move the log on the carriage that transports it past the blade. The basic elements of that technology were known in Japan, but no other machine of comparable complexity was in use, and considerable trial and invention would have been required to develop a sawmill. In addition, the difficulty of log transport prompted lumbermen to do the basic shaping of heavy pieces, for which such a mill would have been most valuable, near the felling site. However, the steeply sloped character of most woodland would have made erection of a mill difficult there, and the erratic character of stream flow would have made construction and operation of a waterwheel equally difficult.

In one way or another, then, the potential advantages of water-powered sawmills were undercut, and that technological development did not occur. Instead, sawyers continued to rip timbers with broad, heavy-duty hand saws, split them with broad-axes and wedges, and shape them with chisels and planes of diverse types.

Whether in the final analysis the lack of lumber wagons, crosscut saws, and water-powered sawmills constituted net gains or losses in overall social or ecological efficiency is probably impossible to say. In the more parochial, short-range terms of lumber provisioning itself, the devices surely would have offered advantages. But even without them, Edo-period lumbering arrangements did provision a large number of exceptionally wood-dependent cities by extracting the yield from uncommonly difficult terrain.

Indeed, the fact that Tokugawa builders—unlike those in so many parts of the world—never shifted from basic dependence on wood construction to mud, clay, or stone may constitute the most striking testimony to the adequacy of these lumbering arrangements. Until the eighteenth century, one could argue, those other building materials had only marginal uses because construction wood was reasonably plentiful and easier to use than alternative materials and provided more pleasant housing. By then architectural canons may have been so well entrenched that, in the absence of more basic challenges to established social norms, other methods of building were simply unthinkable despite intensified lumber scarcities. It is possible, of course, that fear of earthquake damage discouraged stone or other heavy types of construction. However, there seems to be no evidence supporting this proposition, and such fears did not prevent the use of roof tile, which entailed a comparable hazard.

The heavy reliance on wood may betray difficulty in obtaining appropriate materials for alternative types of construction. Japan's complex geological history endowed the archipelago with a broad array of metallic and nonmetallic mineral resources, so that varieties of building stone, clay, and mud and the limestone used in mortared construction were (and are) all present in substantial amounts.[14] However, that geological complexity scattered many of the minerals in small, poorly accessible deposits, and the topography of the islands has made stone quarrying and transport particularly difficult.

Early modern builders used stone to line waterways, face ramparts, form walkways, entomb ashes of the dead, and provide decoration, but not as free-standing walls of buildings. Even where stone was used, builders commonly used surface rocks, such as rounded stones retrieved from riverbeds, for routine work, saving the laboriously shaped quarry stone for corners and other architectural details. Why rock was not used in cemented, free-standing walls is unclear. It may be that because Japan was not overrun by Pleistocene glaciers and lacked naturally clean caches of glacially formed sand and gravel, only the mixed materials of streambeds were available for mortared construction unless one engaged in the tedious tasks of pulverizing, screening, and washing. Alternatively, the absence of such construction may simply reflect ignorance of mortaring techniques despite the availability of requisite ingredients.

More perplexingly, clay was used for tile roofing but not for brick work even though brick became a major building material in China. The absence of brick structures may reflect the comparative scarcity, and consequent preciousness, of usable clay. That preciousness is suggested by the occasional lawsuit that resulted from clay-hungry tile makers damaging paddy fields by digging up the clay pan that made them watertight. It is also suggested by the way potters prized particular sources of clay. Mud (clay with a heavy admixture of coarser particulates) was used extensively in fireproofed warehouses, outdoor perimeter walls, the interior walls of wood-framed buildings, and the undersurfacing of tile roofs. However, adobe housing per se did not exist, perhaps because the high levels of rainfall made it impractical.

Whatever the reason(s) for its predominance, wood construction remained normative until the late nineteenth century, and the lumber industry examined in preceding chapters provided the material for that work. In the process of doing so the industry displayed an overall character and evolution that was basically similar to that of other contemporary industries. The record of early modern lumbering thus illustrates both that broader commercial history and how essential to Tokugawa civilization were the undomesticated natural resources provided by these commercialized hunter-gatherers.

APPENDIX A

Plant Species

The trees and vines mentioned in the text have the following Latin binomials and English equivalents.

JAPANESE	LATIN BINOMIAL	ENGLISH
fujitsuru	*Wisteria floribunda*	a wisteria vine
hinoki	*Chamaecyparis obtusa*	Japanese cypress
kuri	*Castanea crenata*	Japanese chestnut
kurobe (kurobi)	*Thuja standishii*	Standish arbor vitae
matsu	*Pinus* sp.	pine
sawara	*Chamaecyparis pisifera*	Sawara cypress
shirakuchizuru (sarunashi)	*Actinidia arguta* Planch. et Miq.	Stahlengriffel genus
sugi	*Cryptomeria japonica*	cryptomeria

APPENDIX B

Table of Measurements

In the text longer distances are expressed in metric terms: kilometer. However, because the short linear measurements used in Tokugawa Japan were very nearly equal to the comparable English measurements, I have retained the English usages of inch *(sun)* and foot *(shaku)*.

Note: The *ryō* was a gold coin of varying weights and fineness.

The Japanese measurements mentioned in the text have the following metric and English equivalents.

Japanese	Metric	English (or U.S.)
bu	0.375 g	0.013 oz.
kanme (kan) (1,000 *monme*)	3.75 kg	8.72 lbs.
ken (6 *shaku*)	1.82 m	1.99 yds.
koku (grain) (100 *shō*)	180.51 l	5.12 bushels (U.S.)
koku (wood) (1x1x10 *shaku*)	0.27 m^3	9.70 ft.^{3}
monme (10 *bu*)	3.75 g	0.13 oz.
shaku (10 *sun*)	0.30 m	0.99 ft.
shakujime (1x1x12 *shaku*)	0.32 m^3	11.64 ft.^{3}
shō	1.81 l	1.64 qt. (U.S.)
sun	3.03 cm	1.19 in.
tsubo (6x6 *shaku*)	3.30 m^2	3.95 sq. yds.

APPENDIX C

Yoshino *Ringyō:* A Research Topic

Early modern forestry in the mountainous Yoshino district of Yamato Province (Nara Prefecture) has received substantial scholarly attention. It constitutes an excellent area for research because the pertinent primary and secondary materials are comparatively rich, making it a fine vehicle for examining both the historiography and specific topics in the socioeconomic history of forestry during the past four centuries. Scholars of Yoshino forestry have vigorously debated questions about rural-urban relations, landlordism, rich-poor relations, commercial development, the social effects of money, social structure, social values, and social change. Issues relating to technology and the environment are also amenable to study from the perspective of Yoshino's forest history.

A good introduction to the subject is the 1984 review essay by Katō Morihiro, "*Yoshino ringyō zensho* no kenkyū." Katō's point of departure is Mori Shōichirō's pioneering work of 1898, *Yoshino ringyō zensho*, and he proceeds from there to discuss the major works and topics of subsequent scholarship. In particular, he cites the 1962 work by Kasai Kyōetsu, *Yoshino ringyō no hatten kōzō*, which shaped the discourse that followed. Contributors to that discourse, whose works Katō identifies, include Arimoto Sumiyoshi, Fujita Yoshihisa, Fujita Yoshitami, Handa Ryōichi, Iwanaga Yutaka, Izumi Eiji, Tani Yahei, and Yamada Tatsuo. One should also mention the essay by Mitsuhashi Tokio, "Yoshino-

Kumano no ringyō," and another piece by Iwanaga, "Edo-Meiji ki ni okeru Yoshino ringyō no ikurin gijutsu."

The study of Yoshino forestry has continued since Katō wrote: in an essay of 1986, Fujita Yoshihisa examined the impact of plantation forestry on village life, and in three essays of 1987, 1989, and 1990, Izumi Eiji has explored the relations among urban lumbermen, local lumbermen, and less well-to-do villagers as they were evidenced in lumbering work. These four essays have all appeared in *TRK Kenkyū kiyō*. The Bibliographical Essay in this volume offers suggestions on how to search for additional older materials and how to stay abreast of current scholarship.

NOTES

ABBREVIATIONS OF journal names are identified in the Bibliography, which also gives complete citations for items mentioned in the notes.

Notes to Chapter 1

1. For a basic introduction to the Tokugawa economy, see Toyoda Takeshi, *A History of Pre-Meiji Commerce in Japan*, 37–91.

2. This body of economic thought is evident in numerous works. An early and dated but still useful introduction is Honjō Eijirō, *Economic Theory and History of Japan in the Tokugawa Period*, especially chapters 1–5, which are reprinted from essays Honjō published in *KUER* between the late 1920s and early 1940s. Also, Neil Skene Smith, "An Introduction to Some Japanese Economic Writings of the Eighteenth Century." More recent works are Robert N. Bellah, *Tokugawa Religion*; J. R. McEwan, *The Political Writings of Ogyū Sorai*; Tetsuo Najita, *Visions of Virtue in Tokugawa Japan*; and idem, "Political Economism in the Thought of Dazai Shundai (1680–1747)."

3. A lucid and concise examination of this entrepreneurial self-image in the early twentieth century is Byron Marshall, *Nationalism and Capitalism in Prewar Japan*. Najita, *Visions of Virtue*, explores its Tokugawa antecedents, and his essay "Japan's Industrial Revolution in Historical Perspective" suggests linkages in economic thought before and after 1868.

4. This view is skillfully presented in English in Charles David Sheldon, *The Rise of the Merchant Class in Tokugawa Japan, 1600–1868*. For a newer and more concise statement by Sheldon, see his " 'Pre-Modern' Merchants and Modernization in Japan." A prewar work is Honjō Eijirō, *The Social and Economic History of Japan*, an assemblage of essays first published in *KUER*. In chapter 8, Honjō highlights the oppressive condition of peasant life under feudal rule. A similar perception is evident in Hugh Borton, "Peasant Uprisings in Japan of the Tokugawa Period."

5. An insightful recent study of Marxist thought in Japan is Germaine Hoston, *Marxism and the Crisis of Development in Prewar Japan*. As Hoston skillfully shows, Marxian analysis could generate vigorous disagreement, and considerable "Marxian" study of Tokugawa history was Marxian only in a general sense. Honjō's work, mostly during the 1930s, hardly meets a strict Marxian definition; nor does the early postwar writing of Horie Yasuzo, as evidenced in his essay "The Feudal States and the Commercial Society in the Tokugawa Period." The task of countering Western supremacism also seems operative in the creation and use by Japanese intellectuals of the concept of *tōyōshi* or "East Asian history," as recently examined in Stefan Tanaka, *Japan's Orient*. Postwar approaches to history, mainly of the twentieth century, are examined more broadly by Carol Gluck in "The Past in the Present."

6. The fascist celebration of forceful, charismatic leadership is reflected in the Preface of Arthur L. Sadler, *The Maker of Modern Japan: The Life of Tokugawa Ieyasu*. A Marxian approach to class relationships and historical dynamics is expressed in the pioneering works of E. H. Norman, notably his influential *Japan's Emergence as a Modern State*. On Norman and his scholarship, see John W. Dower, ed., *Origins of the Modern Japanese State: Selected Writings of E. H. Norman*; Roger Bowen, ed., *E. H. Norman, His Life and Scholarship*; and additional sources cited therein. The generally negative assessment of Tokugawa history shows up clearly in G. B. Sansom, *Japan, A Short Cultural History*.

7. The most dedicated expression of this latter-day Marxian perspective in Tokugawa studies is Herbert P. Bix, *Peasant Protest in Japan, 1590–1884*.

8. A concise and influential formulation is E. S. Crawcour, "Changes in Japanese Commerce in the Tokugawa Period." A monographic study that illustrates this view as it applies to cotton production in the Osaka region is William B. Hauser, *Economic Institutional Change in Tokugawa Japan*. For the most influential interpretation of the rural ramifications of early modern commercial change, see Thomas C. Smith's classic, *The Agrarian Origins of Modern Japan* and select essays in his *Native Sources of Japanese Industrialization, 1750–1920*.

9. A concise expression of this view is Crawcour, "The Premodern Economy." See also the essay by John Whitney Hall, "The Nature of Traditional Society: Japan"; and two by Crawcour, "The Tokugawa Heritage" and "The Tokugawa Period and Japan's Preparation for Modern Economic Growth."

This view found its most ambitious formulation during the later 1960s in a set of five volumes of essays known collectively as the Princeton *Studies in the Modernization of Japan*. The essays in the Princeton series cover a wide range of topics, mostly on post-1868 history, and not all are consciously engaged in promoting the modernizationist position. The core logic of the series is discussed in Hall, "Changing Conceptions of the Modernization of Japan," the opening essay in Marius B. Jansen, ed., *Changing Japanese Attitudes Toward Modernization*, the first volume in the set. For a critical examination of the ideology that informed this series, see H. D. Harootunian, "America's Japan/Japan's Japan."

The modernizationist perspective has been particularly evident in the rich scholarly output of Susan B. Hanley and Kozo Yamamura, most notably in their *Economic and Demographic Change in Preindustrial Japan, 1600–1868*. For an overall formulation, see Yamamura, "Toward a Reexamination of the Economic History of Tokugawa Japan, 1600–1867." A skillful application of this approach to the seventeenth century is his "Returns on Unification: Economic Growth in Japan, 1550–1650." On the later Edo period, see Hanley, "A High Standard of Living in Nineteenth-Century Japan: Fact or Fantasy?"; her exchange with Yasuba Yasukichi in *JEH* 46, 1 (Mar 1986): 217–226; and her "How Well Did the Japanese Live in the Tokugawa Period? A Historian's Reappraisal." A conference volume that reflects this approach is Erich Pauer, ed., *Silkworms, Oil, and Chips: Proceedings of the Economics and Economic History Section of the Fourth International Conference on Japanese Studies*. A more recent work in this vein is Hanley, "Tokugawa Society: Material Culture, Standard of Living, and Life-styles."

10. This interpretive shift is discussed approvingly in Susan B. Hanley and Kozo Yamamura, "A Quiet Transformation in Tokugawa Economic History." For examples in English, see Akimoto Hiroya, "Capital Formation and Economic Growth in Mid-nineteenth-Century Japan"; Nishikawa Shunsaku, "Productivity, Subsistence, and By-employment in the Mid-nineteenth-Century Chōshū"; and more recently Nakamura Satoru, "The Development of Rural Industry." See also essays in Hayami Akira, ed., *Pre-conditions to Industrialization in Japan,* several of which were published in *ESQ,* including the Hanley essay cited above; Yasuba Yasukichi, "The Tokugawa Legacy: A Survey"; and Nishikawa, "The Economy of Chōshū on the Eve of Industrialization."

11. These premises undergirded an approach to political history in which authors contended that Japan's success in modernizing could be ascribed to its "feudal" experience. Europe, this argument held, had been feudal (by a non-Marxian definition) and had subsequently modernized, and elsewhere only Japan had a "true" feudal experience, which background gave it a set of values and human relationships that enabled it to modernize. This theme was most fully articulated in essays in Rushton Coulborn, *Feudalism in History.* See John Whitney Hall, "Feudalism in Japan—A Reassessment," for a discussion of the concept of feudalism, the choice of a particular definition, and its application to Japanese elite political history. A concise narrative application of this perspective is Peter Duus, *Feudalism in Japan.*

12. On early Tokugawa merchant-government collaboration, see Crawcour, "Changes in Japanese Commerce." For a more richly detailed example, see Crawcour, "Kawamura Zuiken: A Seventeenth-Century Entrepreneur." Also, James L. McClain, "Castle Towns and Daimyo Authority: Kanazawa in the Years 1583–1630."

For the later Edo period, information on ruler-merchant collaboration is scattered. William Kelly, *Deference and Defiance in Nineteenth-Century Japan,* notes the role of the merchant Honma in Shōnai domain. He cites the Ph.D. dissertation of Linda L. Johnson, "Patronage and Privilege: The Politics of Provincial Capitalism in Early Modern Japan." Johnson adumbrates the matter in "Prosperity and Welfare: The Homma Family and Agricultural Improvement in Meiji Japan." Bix, *Peasant Protest,* describes merchant-ruler ties in Fukuyama. The doctoral dissertations of Mark J. Ravina and Luke S. Roberts, "Political Economy and Statecraft in Early Modern Japan" (Stanford, 1991) and "The Merchant Origins of National Prosperity Thought in Eighteenth-Century Tosa" (Princeton, 1991) respectively, reveal much more fully the linkages of government and merchants in daimyo domains, as does the work of Sarah Metzger-Court, foreshadowed in a brief essay in Gordon Daniels, ed., *Europe Interprets Japan.* Conrad Totman, *The Green Archipelago,* notes the linkages of government, village, and entrepreneur in late Tokugawa reforestation projects.

13. A vigorous statement is Seymour Broadbridge, "Economic and Social Trends in Tokugawa Japan." English-language works on demography, social unrest, and social hardship are numerous. For a fairly complete listing of books and articles published before 1985, see the bibliographical essay by Conrad Totman, "Tokugawa Peasants: Win, Lose, or Draw?" A more recent essay on rural hardship is David L. Howell, "Hard Times in the Kantō: Economic Change and Village Life in Late Tokugawa Japan."

14. See, for example, the nicely turned essay by Tessa Morris-Suzuki, "Concepts of Nature and Technology in Pre-Industrial Japan."

15. There are already, of course, well-developed fields of environmental studies in the realms of both Japanese- and English-language scholarship as a whole. However, the application of environmental perspectives to Tokugawa studies is only beginning, as suggested by the pioneering character of Andō Seiichi's new study of environmental pollution, *Kinsei kōgaishi no kenkyū.* Among older works on Tokugawa history, two merit note as sophisticated, pioneering ecological studies: Iwasaki Naoto's grand study of Akita forests, *Akita ken Noshirogawa kami chihō ni okeru sugibayashi no seiritsu narabi ni kōshin ni kan suru kenkyū,* and Chiba Tokuji's study of denuded mountains, *Hageyama no kenkyū.* Neither Iwasaki nor Chiba was an academic historian.

16. This topic has many dimensions. One, for example, is the role of microbes, the etiological agents of communicable disease. A fine pioneer study is Ann Bowman Jannetta, *Epidemics and Mortality in Early Modern Japan.* Another dimension is environmental pollution. Still unstudied in English, Andō's new book, *Kinsei kōgaishi no kenkyū,* is the pioneer work in this field. Awareness of geography and of its influence on history, which is central to this approach, is the topic of Kären Wigen, "The Geographic Imagination in Early Modern Japanese History: Retrospect and Prospect." See also her book-length study of Tokugawa-Meiji regional development, *The Making of a Japanese Periphery, 1750–1920.* For a general treatment and reference to further works on aspects of this environmental history, see Conrad Totman, *Early Modern Japan,* chapters 11–13.

17. On the general role of government in the early Tokugawa economy, see Wakita Osamu, "The Social and Economic Consequences of Unification," and Nakai Nobuhiko, "Commercial Change and Urban Growth in Early Modern Japan." Other works that show government-led construction work more particularly are Mary Elizabeth Berry, *Hideyoshi;* William B. Hauser, "Osaka Castle and Tokugawa Authority in Western Japan"; James L. McClain, *Kanazawa;* and Conrad Totman, *Tokugawa Ieyasu, Shogun.*

18. On mining, see Robert Leroy Innes, *The Door Ajar: Japan's Foreign Trade in the Seventeenth Century,* and Nagahara Keiji and Kozo Yamamura, "Shaping the Process of Unification: Technological Progress in Sixteenth- and Seventeenth-Century Japan."

The monetary system has received much attention; for example, Kozo Yamamura and Kamiki Tetsuo, "Silver Mines and Sung Coins—A Monetary History of Medieval and [Early] Modern Japan in International Perspective." See also relevant essays in the special issue of AA 39 (1980), "Studies in the History of Japanese Currency Systems." Other valuable titles can be found in William D. Wray, *Japan's Economy: A Bibliography of Its Past and Present,* 11–12. The monetary system was given greater flexibility, and problems of capital scarcity ameliorated, by the development of credit instruments that facilitated the pooling and redistribution of funds. A study that examines credit mechanisms of the early Edo period is E. S. Crawcour, "The Development of a Credit System in Seventeenth-Century Japan." Two that deal with late Edo are Arne Kalland, "A Credit Institution in Tokugawa Period," and Ronald P. Toby, "Both a Borrower and a Lender Be: From Village Moneylender to Rural Banker in the Tempō Era."

19. On foreign trade in this regard, see in particular Innes, *The Door Ajar.* For other titles on foreign trade, see the excellent listing in Wray, *Japan's Economy: A Bibliography,* 23–26. On seventeenth-century merchants more generally, see Crawcour, "Changes in Japanese Commerce"; items cited in note 22 below; and Wray, *Japan's Economy: A Bibliography,* 18–21.

20. A general study of urban development that contains considerable material on castle towns is Yazaki Takeo, *Social Change and the City in Japan.* McClain, *Kanazawa,* is an excellent monographic study. Also see Hall, "The Castle Town and Japan's Modern Urbanization."

21. On merchant organization in general, see Crawcour, "Changes in Japanese Commerce"; Sheldon, *The Rise of the Merchant Class;* and Hauser, *Economic Institutional Change.* A dated but useful introduction to *nakama* and their antecedents is Toyoda Takeshi, "Japanese Guilds." More specialized treatments are W. Mark Fruin, "The Firm as Family and the Family as Firm in Japan"; Takatera Sadao and Nishikawa Noboru, "Genesis of Divisional Management and Accounting Systems in the House of Mitsui, 1710–1730"; and Irie Hiroshi, "Apprenticeship Training in Tokugawa Japan." A recent treatment of merchant households is Sakudō Yōtarō, "The Management Practices of Family Business."

22. On merchant codes, see Crawcour, trans., "Some Observations on Merchants," which is a translation of Mitsui Takafusa's *Chōnin kokenron,* a

long exegesis on merchant values and conduct. Also, J. Mark Ramseyer, "Thrift and Diligence: House Codes of Tokugawa Merchant Families."

23. See Yazaki, *Social Change and the City*, in particular, and also Gilbert Rozman, *Urban Networks in Ch'ing China and Tokugawa Japan.*

24. For a concise treatment of Osaka, see William B. Hauser, "Osaka: A Commercial City in Tokugawa Japan." Older essays that focus on Osaka are Miyamoto Mataji, "Economic and Social Development of Osaka," and idem, "The Merchants of Osaka." See also Mori Yasuhiro, "Loans to Daimyos by the Osaka Money Changers," and Crawcour, "The Development of a Credit System in Seventeenth-Century Japan."

25. A pioneering study of *sankin kōtai* that became the classic statement in English is Toshio G. Tsukahira, *Feudal Control in Tokugawa Japan.* A new study of the highway system that expedited *sankin kōtai* travel is Constantine N. Vaporis, "Overland Communications in Tokugawa Japan." Also idem, "Post Station and Assisting Villages: Corvée Labor and Peasant Contention." Wigen, *The Making of a Japanese Periphery, 1750–1920,* also examines overland transport. On the corollary development of a communications network, see Moriya Katsuhisa, "Urban Networks and Information Networks."

26. Despite its historical importance, the "reform-minded" regime of the shogun Yoshimune has yet to be adequately treated in English. Aspects of his commercial policy are treated in Hauser, *Economic Institutional Change;* Tsuji Tatsuya, "Politics in the Eighteenth Century"; and Totman, *Early Modern Japan,* chap. 14.

27. See Jannetta, *Epidemics and Mortality,* on the general problem of infectious disease. On urban demography, see the essays by Hayami Akira and Sasaki Yōichirō in Susan B. Hanley and Arthur P. Wolf, eds., *Family and Population in East Asian History.* Also Robert J. Smith, "Small Families, Small Households, and Residential Instability: Town and City in Pre-Modern Japan," and idem, "Aspects of Mobility in Pre-Industrial Japanese Cities." A new, richly detailed study of diverse urban workers is Gary Leupp, *Servants, Shophands, and Laborers in the Cities of Tokugawa Japan.*

28. Ground coal and fish meal production have yet to receive monographic treatment in English. However, the early modern fishing industry more broadly is beginning to come under scrutiny. The anthropologist Arne Kalland has published articles on Tokugawa fisheries, mainly those of northwest Kyushu see his *Fishing Villages in Tokugawa Japan.* David L. Howell, "Proto-Industrial Origins of Japanese Capitalism," and his forthcoming book, *Indigenous Capitalism in Nineteenth-Century Japan,* treat fishing in northern Japan. Also, some of the essays in Kenneth Ruddle and Akimichi Tomoya, eds., *Maritime Institutions in the Western Pacific,* touch on Japan's preindustrial fisheries.

Salt production often was associated with fishing villages, and it is treated in two old studies, John D. Eyre, "Patterns of Japanese Salt Production and Trade," and A. E. Wileman, "Salt Manufacture in Japan." Another related topic that still invites full-scale attention is the coasting trade, which is treated in Crawcour, "Kawamura Zuiken" and "Notes on Shipping and

Trade in Japan and the Ryukyus"; in essays of Robert G. Flershem, most notably "Some Aspects of Japan Sea Shipping and Trade in the Tokugawa Period, 1603–1867"; and in a set of essays in JAS 23, 3 (May 1964).

On plantation forestry, see Totman, *Green Archipelago*. On foreign trade in silk, see Innes, *The Door Ajar*. The rise of a domestic silk floss industry has not received extended treatment, but the practice of sericulture is described in Stephen Vlastos, *Peasant Protests and Uprisings in Tokugawa Japan*, 92–113; and Tessa Morris-Suzuki, "Sericulture and the Origins of Japanese Industrialization," discusses technical aspects that contributed to Japanese silk's later export potential. William J. Chambliss, *Chiaraijima Village*, 17–25, examines its role in one locality of the Kantō plain.

29. Thomas C. Smith has examined these agronomic developments most fully, notably in his *Agrarian Origins*, in which he advances a "developmental economics" explanation for this rural commercial activity. See also idem, "Ōkura Nagatsune and the Technologists." On trends in irrigation, a fine case study is William W. Kelly, *Water Control in Tokugawa Japan*. For a recent treatment of the Tokugawa rural scene, see Satō Tsuneo, "Tokugawa Villages and Agriculture," and Furushima Toshio, "The Village and Agriculture during the Edo Period."

30. On the use of urban refuse as fertilizer, see Anne Walthall, "Village Networks: Sōdai and the Sale of Edo Nightsoil," and Susan Hanley, "Urban Sanitation in Preindustrial Japan." On fish meal, see the titles on fisheries cited in note 28 above.

31. On cash income, see particularly T. C. Smith, "Farm Family By-Employments in Preindustrial Japan." Concerning work away from home *(dekasegi)*, see the several essays on migration by W. Mark Fruin and others cited in Wray, *Japan's Economy: A Bibliography*, 16–17. For econometric analyses of late Tokugawa rural wage trends, see Nishikawa Shunsaku's two essays cited in note 10 above; his "Protoindustrialization in the Domain of Chōshū in the Eighteenth and Nineteenth Centuries"; and also three essays by Saitō Osamu: "Population and the Peasant Economy in Proto-Industrial Japan," "The Labor Market in Tokugawa Japan: Wage Differentials and the Real Wage Level, 1727–1830," and "Changing Structure of Urban Employment and Its Effects on Migration Patterns in Eighteenth- and Nineteenth-Century Japan."

32. On literacy and learning, see Ronald P. Dore, *Education in Tokugawa Japan*, and Richard Rubinger, *Private Academies of Tokugawa Japan*. On rural arts and letters in particular, see Anne Walthall, "Peripheries: Rural Culture in Tokugawa Japan." T. C. Smith's writings treat the question of rural literacy and learning, and their economic implications are addressed in J. I. Nakamura, "Human Capital Accumulation in Premodern Rural Japan."

33. Studies that examine the rise of rural entrepreneurs include Sheldon, *The Rise of the Merchant Class*; Chambliss, *Chiaraijima Village*; Vlastos, *Peasant Protests and Uprisings*; and Hauser, *Economic Institutional Change*. More recent treatments are Nakamura Satoru, "The Development of Rural Industry," and the doctoral thesis by Edward Pratt, "Village Elites

in Tokugawa Japan." On the Kinai cotton industry specifically, see also Hauser, "The Diffusion of Cotton Processing and Trade in the Kinai Region in Tokugawa Japan," and Shimbo Hiroshi, "A Study of the Growth of Cotton Production for the Market in the Tokugawa Era." The activities of rural merchants have recently been fitted into the concept of "protoindustrialization," which is a refinement of modernization theory. Early statements are Nishikawa, "Protoindustrialization in the Domain of Chōshū," and Saitō Osamu, "Population and the Peasant Economy." For more recent treatments, see Howell, "Proto-Industrial Origins of Japanese Capitalism"; Toby, "Both a Borrower and a Lender Be"; and Wigen, *The Making of a Japanese Periphery, 1750–1920.*

34. Fiscal maneuvers by the rulers have been extensively studied. See entries cited in Wray, *Japan's Economy: A Bibliography,* 11. On stipend reduction in particular, see Kozo Yamamura, "The Increasing Poverty of the Samurai in Tokugawa Japan, 1600–1868." See also J. Victor Koschmann, *The Mito Ideology,* for information on Mito finances in the later Edo period. Katsu Kokichi, *Musui's Story,* and Yamakawa Kikue, *Women of the Mito Domain,* offer glimpses of the fiscal circumstances of middle-ranking samurai households.

Notes to Chapter 2

1. William H. Coaldrake, "Edo Architecture and Tokugawa Law," conveys nicely a sense of architectural decline, even in the politically advantaged city of Edo.

2. Nishikawa Zensuke, "Ringyō keizai shiron (1)," 13, subdivides Tokugawa lumbering into five categories. My first category—subsistence provisioning—is external to his five. I combine his 1 and 2 in my second (command) category and collapse his 3, 4, and 5 into my third (entrepreneurial). Nishikawa's categories are as follows:

> 1. *chūseiteki mokuzai seisan:* medieval-style lumber production, in which hereditary assigned loggers extracted and transported timber on behalf of the estate holder for whom they worked
> 2. *ryōshūteki mokuzai seisan:* lumber production by fief-holding lords, in which lords obtained timber from forests on their fiefs by having subordinates supervise corvée labor and skilled loggers, the output being regarded as tax payments or legitimate feudal obligations
> 3. *shōminteki mokuzai seisan:* merchant lumber production, in which small-scale merchants, who ideally were semipeasant villagers, arranged the provisioning as a way to supplement their agricultural income
> 4. *koshōhin seisan zaimoku seisan:* production of small-size wood products, essentially handicraft work by local producers
> 5. *shihonka mokuzai seisan:* capitalist lumber production, in which logging was capitalized and controlled by large-scale entrepreneurs who ideally were urbanites, outsiders to the logging area, and who directly or indirectly used their capital to control the woodland and its occupants on behalf of their own interests

3. Although fuelwood and charcoal provisioning lie beyond the scope of this study, those goods were handled in much the same way as lumber. It is important to remember that there probably was more wood cut for fuel than for lumber even though most homes were heated minimally. And because the poor used nearly as much fuel as the rich, fuel wood was cut mostly by subsistence woodsmen. It should also be noted, as is stressed in Katō Morihiro, "Edo chimawari sanson no gōnō keiei," that entrepreneurial charcoal, firewood, and lumber provisioning were clearly linked in many areas, with the fuel supplies that were borne by log rafts constituting a crucial part of the provisioner's income.

4. The richness of documentation on command lumbering reflects the care with which the Meiji government assembled Edo-period records of *bakufu* and *han* forest management. The topic is discussed in the Bibliographical Essay of Totman, *Green Archipelago*. Initially scholarship focused on command lumbering, but entrepreneurial activity has now become the main area of inquiry.

5. Tokoro Mitsuo, "Ringyō," 205. Kyōgoku's wood was *kigawara, itago,* and *kureki* (wooden shingles, planking, and split pieces); Akita's was semifinished timbers. To a degree the difference in goods reflects differences in stand composition, but mainly it reflects differences in transport capacity: the Tenryū could not handle large timbers at that time whereas the Omono and Yoneshiro could.

6. Shioya Junji, "Fushimi chikujō to Akita sugi," 49–51. Tokoro Mitsuo, *Kinsei ringyōshi no kenkyū,* 23–31. Shioya discusses the problem of determining exactly what Hideyoshi's measurements meant, noting that his orders became increasingly precise and standardized. The analysis suggests that Hideyoshi acquired a keen sense of the difficulty loggers faced in converting trees to regularly sized pieces of lumber.

7. Tokoro Mitsuo, "Ieyasu kurairichi jidai no Kiso kanjō shiryō," 329–330. The sticks of shingling *(doi)* from Shinano appear to have measured about three feet long by nine inches wide. Split into shingles at the production site, they went out by pack horse, 60 per load if thick shingles, 90 per load if thin. The recorded numbers of pieces—47,668; 943; 5,996—which are the numbers of serviceable sticks actually received, suggest that Ieyasu ordered in round numbers (50,000, 1,000, 6,000).

8. Tokoro, "Ieyasu," 329. The Kiso shingling *(kawaragi)* was "large" pieces of *hinoki,* which may have meant pieces about four feet in length and roughly five by seven inches face measure.

9. Hirao Michio, *Tosa han ringyō keizaishi,* 164–165.

10. Information on Numata is from Ōsaki Rokurō, "Ohayashi chiseki kakutei katei no ichi kenshō," 20–21. The trees were *kurobe,* and the volume of wood was to equal eight thousand timbers approximately twelve feet long by one foot square.

11. The handling of tax *(nengu)* timber produced by villages in forested areas of *bakufu* domain in the Ina district of the Tenryū valley in Shinano has been particularly well studied by Iioka Masatake in essays in *TRK Kenkyū kiyō* 47, 50, 52, and 61 (Mar 1973, 1976, 1978, 1987).

12. Tokoro, *Kinsei ringyō,* together with a number of essays he did not reproduce or revise for that volume, examines the Kiso case in detail. For a listing of those essays, see Totman, *Green Archipelago,* 285–286. For an introduction to Akita lumbering and guidance to the relevant scholarly literature, see Conrad Totman, *The Origins of Japan's Modern Forests: The Case of Akita.*

13. From Michishige Tetsuo, "Hansei goki ni okeru hanyō mokuzai no seisan kōzō," 2–5.

14. Shioya Tsutomu, *Buwakebayashi seido no shiteki kenkyū,* 56–57.

15. Shimada Kinzō, "Bakumatsu no goyōzai shidashibito Shinanoya Shōzaburō no gyōtai (I)," 62–63.

16. *Daijinmei jiten,* 2:297. For monographic detail on Kinokuniya's logging work on the Ōi river west of Mt. Fuji in the 1690s, see Shimada Kinzō, "Bakumatsu no goyōzai shidashibito Shinanoya Shōzaburō no gyōtai (II)," 50–59. Kinokuniya's celebrated rise sounds suspiciously like that attributed to Kawamura Zuiken (1618–1699), on which see Crawcour, "Kawamura Zuiken," 28–50.

17. Information on Nogami is from Fujita Yoshitami, *Kinsei mokuzai ryūtsūshi no kenkyū,* 212–225. Converting to English units, I estimate Nogami's 1.13 *koku* of 1783 to be about one-third acre of arable land. Fujita, *Kinsei mokuzai,* 217, reports that in 1838, Rihyōe possessed about 2.9 acres of arable and 3.7 of woodland. In 1864 he held 3.2 acres of arable and 6.1 of forest.

18. Information from Shimada, "Bakumatsu (I)," 60, 65–67, 77–80. For monographic detail on Shinanoya's logging, mainly on the Ōi River west of Mt. Fuji, see Shimada, "Bakumatsu (II)" and "Bakumatsu (III)." The latter piece also contains an addendum that furnishes genealogical detail on the Noguchi family.

19. Shimada, "Bakumatsu (I)," 77; Tokoro, *Kinsei ringyō,* 298. The official abolition of lumbermen's guilds in 1842 temporarily disrupted but did not effectively dissolve the wholesaler-broker procedures that had been basic to Edo lumbering since the seventeenth century. The *kabu nakama* arrangements of lumbermen were officially recognized again in 1851, but with a substantially changed membership. Shimada, "Bakuhan taiseika no Edo zaimokushō no shōtai," 160–162; and Shimada, "Bakuhan taiseika no Edo zaimokushō no shōtai II," 98–99.

20. Shimada Kinzō, "Kawabe ichibangumi koton'ya kumiai monjo to Edo zaimoku ichiba," 53. The "three generations" presumably were the diligent, frugal founder, his profligate, ne'er-do-well son; and his impoverished grandson.

21. Shimada, "Bakuhan taiseika . . . II," 95–99.

22. On the development of regenerative forestry—tree planting—as a major source of new timber, see Totman, *Green Archipelago,* and, for the Akita region specifically, idem, *Origins.*

23. This description of financing is based on material scattered through several works, most notably Shimada Kinzō, *Edo-Tōkyō zaimoku ton'ya kumiai seishi.*

24. The lumber business was further complicated by the trimetal monetary system and regional and temporal fluctuations in relative money values. *Ton'ya* played key roles in converting money so that transactions could be carried out. See Shimada, *Edo-Tōkyō*, 574–575.

25. Yoshida Yoshiaki, *Kiba no rekishi*, 257–262. The transaction date is not given. Most of the lumber was *sugi* and *hinoki* board stock and split pieces.

26. For a description of modern forest mensuration techniques, see Grant W. Sharpe et al., *Introduction to Forestry*, 251–280.

27. Iioka Masatake examines this case in "Kinsei chūki no yōzai seisan shihō to saiunhi," 107–121.

28. Nishikawa Zensuke, "Edo zaimokushō no kigen: Edo mokuzai ichibashi josetsu," 7, 13–16.

29. Shimada, *Edo-Tōkyō*, 586–587.

30. Ibid., 529–532, and Nishikawa, "Edo zaimokushō," 8–16, describe changes in lumberyard locations. Besides entrepreneurial lumberyards, the *bakufu* and major timber-producing *han* maintained their own yards for storing government wood.

31. Shimada, *Edo-Tōkyō*, 536.

32. Sunaga Akira, "Kinugawa jōryūiki ni okeru ringyō chitai no keisei," 13. For the later history of these Kinugawa lumbermen, see Abe (erstwhile Sunaga) Akira, "Kinsei goki Kita Kantō ni okeru ringyō no hatten," in which the author discusses the linkage between these lumbering costs and the intensified social stratification that accompanied the rise of entrepreneurial lumbering in late Tokugawa villages in the northern Kantō. See also Abe, "Kinsei ni okeru Kinugawa jōryūiki no ikada nagashi."

33. Shimada, *Edo-Tōkyō*, 538. Shimada adds detail to our understanding of the Edo lumber market in essays published in *TRK Kenkyū kiyō* 61 (Mar 1987) and the newly renumbered vols. 23, 24, 26 (Mar 1989, 1990, 1992). Only his death at age 88 in 1992 halted Shimada's scholarly output.

34. Shimada, *Edo-Tōkyō*, 537, 540–541, 549. Idem, "Kawabe," 45. Master carpenters performed the functions of building contractors.

35. Shimada Kinzō, "Bakuhan kenryoku kōzōka no zaimoku ton'ya nakama no kōdō," 23–28.

36. Shimada's writings mostly focus on these wholesalers.

37. Iioka Masatake's articles on Tenryū logging, cited in notes 11 and 27 above, detail complexities in the rural end of this provisioning. Yoshida, *Kiba no rekishi*, 242–244, outlines the overall process briefly. Shimada, "Kawabe," 54–55, and *Edo-Tōkyō*, 561, comment on shipping by sea.

38. Yoshida, *Kiba no rekishi*, 244–250. Three good studies of Naguri lumbering that examine its social character and implications have appeared in *TRK Kenkyū kiyō*. They are by Katō Morihiro (56 [Mar 1982] and 61 [Mar 1987]) and Wakino Hiroshi (59 [Mar 1985]). For logging and rafting on the Tama River, see Matsumura Yasukazu, "Ōme ringyō ni okeru ikada," and idem, "Ōme no ringyō."

39. Shimada, *Edo-Tōkyō*, 538–541, 559, 564, 566. Matsumura Yasukazu, "Kinsei Ōme ringyō no seiritsu oyobi hatten ni kan suru rekishi

chirigakuteki kenkyū (shōroku)," 15. Of the several subgroups of the Kawabe *zaimoku ton'ya* group, the Original First Riverside Wholesalers' Guild (Kawabe ichibangumi koton'ya kumiai) was the largest and handled most of the lumber. The other Kawabe *kumi* mostly dealt in products other than timber.

40. Shimada Kinzō, "Edo zaimoku ton'ya no funsō to futan kuyaku," examines the disorder and mobility among *ton'ya*.

41. In Osaka, unlike Edo, commodity specialization seems to have occurred among brokers, and the wholesaler-broker relationship appears to have been more competitive, with brokers boycotting wholesalers who angered them. Indeed, whereas Edo brokers regularly paid commissions to wholesalers, in Osaka wholesalers provided *nakagai* with rebates. Shimada, "Bakuhan kenryoku," 41; Yoshida, *Kiba no rekishi*, 145–147; Funakoshi Shōji, *Nihon ringyō hattenshi*, 48. See also the lucid article by Ōta Katsuya, "Kinsei chūki no Ōsaka zaimoku ichiba."

42. Shimada, "Kawabe," 46–47. Yoshida, *Kiba no rekishi*, 145–147, 153–154.

43. Shimada, "Bakuhan taiseika . . . II," 85.

44. Shimada, *Edo-Tōkyō*, 561, 571.

Notes to Chapter 3

1. Tokoro, *Kinsei ringyōshi no kenkyū*, 16–18; Nihon Gakushiin, *Meiji zen Nihon ringyō gijutsu hattatsushi*, 421; Yamamoto Hikaru, "Azuchi Momoyama jidai no ringyō," 24.

2. Mizutani Seizō, "Kinsei Kyōto no mokuzai yusō (jō)," 43–45.

3. Nihon Gakushiin, 315.

4. Ibid., 308.

5. Ibid., 309, 435; Yamamoto, 24.

6. The Kiso emptied into the Nagara just above Sunomata before it shifted to a new southerly route during a disastrous flood in 1586.

7. Tokoro Mitsuo, "Nishikori tsunaba ni tsuite—kuchie kaisetsu," 102–103; idem, "Unzai chūkei kichi to shite no Inuyama," 2–5.

8. An introduction to the problems of river control is Conrad Totman, "Preindustrial River Conservancy [in Japan]: Causes and Consequences."

9. See Kelly, *Water Control in Tokugawa Japan*.

10. For information on tree species and their distribution, a reliable dendrological work is Kitamura Shirō and Okamoto Shōgo, *Genshoku Nihon jumoku zukan*. A partial listing in English is *Important Trees of Japan*. The latter work is limited to commercially significant trees. The pioneer European descriptive botanical classification for Japan, done in Latin, is Carl Peter Thunberg, *Flora Japonica*.

11. Tokoro Mitsuo, "Suminokura Yōichi to Kisoyama," 8.

12. Nishikawa Zensuke, "Ryūtsū ichiba kara mita mokuzai shōhin seisan no hatten," 6.

13. This description of upper Ki river work is from Mitsuhashi Tokio, "Yoshino-Kumano no ringyō." This Yoshino River, so called because it rises deep in the Yoshino mountains of the Kii peninsula, should not be confused with the larger river of the same name in Shikoku.

14. Matsumura, "Ōme ringyō ni okeru ikada," 19.

15. Hirao, *Tosa han ringyō keizaishi,* 20–25.

16. Iioka Masatake, "Enshū Funagira ni okeru bakufu no kureki shobun," 117–118; and idem, "Enshū Funagira ni okeru bakufu yōzai no chūkei kinō," 116.

17. Hirao, 20–25.

18. Matsumura, "Ōme ringyō ni okeru ikada," 22.

19. Takase Tamotsu, "Kaga han rinsei no seiritsu ni tsuite," 522. The quotation is from 539.

20. Matsumura, "Ōme ringyō ni okeru ikada," 21–22.

21. Nishikawa, "Ryūtsū ichiba kara mita mokuzai shōhin seisan no hatten," 6.

22. Tokoro, *Kinsei ringyō,* 274.

23. Wakino Hiroshi, "Kinsei Nishikawa ringyō ni okeru zaimokushō keiei," 245.

24. Kiso forestry has been extensively studied, most notably in Tokugawa Yoshichika, *Kisoyama,* and the many works of Tokoro Mitsuo, for which see chapter 2, note 12. For this brief summation I have relied heavily on the concise description in Nihon Gakushiin, 296–314.

25. Nihon Gakushiin, 296–298; Tokoro, "Nishikori tsunaba ni tsuite," 116. These two sources differ on some specifics.

26. Split pieces were half, quarter, sixth, or eighth round, depending on the size of the tree, and they were of varying lengths, depending on intended use.

27. Tokoro, *Kinsei ringyō,* 253.

28. Shimada, "Bakumatsu no goyōzai shidashibito Shinanoya Shōzaburō no gyōtai (II)," 36, 41–42, 60.

29. Nihon Gakushiin, 430–432, describes several types of rafts briefly.

30. The material in this section is condensed from ibid., 365–366, 375–419, except as otherwise noted.

31. The word *log* is to be understood in this discussion as embracing semiprocessed pieces, most notably split pieces, which were produced in great quantity (cf. note 36 below).

32. *Ōkawagari* was also called *kudanagashi* or *baragari.* Nihon Gakushiin, 418–419, reports that logs began reaching Nishikori about seven days after the floating season began.

33. Tokoro, "Nishikori," 101.

34. This section on the Nishikori boom and storage area is based on material in ibid., 104–108, 110–111, 114; Nihon Gakushiin, 415–417.

35. Nihon Gakushiin, 415, reports that the rope was 210–220 *ken* (1,254–1,313 feet) in length, but a portion of that was used for tying around the end posts and rock. More modestly, Toba Masao, *Nihon ringyōshi,* 124, reports the rope as some 90 *ken* (537 feet) in length, of which only 60 *ken* were over water.

36. Tokoro, "Nishikori," 104, reports that 60 percent of the 680,000 pieces that passed Nishikori annually during the 1730s and 1740s was *kureki,* a large type of split wood.

37. This section on rafting is based on information in Tokoro,

"Unzai," 2, 21; idem, "Nishikori," 111; and Nihon Gakushiin, 428–429, 433.

38. Tokoro, "Unzai," 2, reports that thirty to fifty rafts were lashed together; idem, "Nishikori," 112, and Nihon Gakushiin (which appears to rely heavily on Tokoro's old Nishikori study), 433, report that forty-eight to fifty-six rafts were combined.

39. This section on marketing and shipping uses material from Nihon Gakushiin, 445–447, 451–455.

Notes to Chapter 4

1. There is a considerable body of scholarship on Yamaguni lumbering. I have examined three books and about a dozen essays, most importantly works by Fujita Yoshitami, Mizutani Seizō, and Motoyoshi Rurio. For a brief review of the scholarship, see Motoyoshi, *Senshin ringyō chitai no shiteki kenkyū,* 15–31. In articles that appeared during the 1980s in successive issues of *TRK Kenkyū kiyō,* Motoyoshi pressed his inquiry into other areas of early modern forestry in the mountains north of Kyoto.

The central concern of these scholars is to analyze the changing power relationships of the groups engaged in this lumbering activity, in broad terms to explore how capitalist lumbering arose and what it meant in terms of rural-urban relationships and social stratification, land control, and power relationships within village society. The book-length work by Fujita, *Kinsei mokuzai ryūtsūshi no kenkyū,* is the fullest examination of the Yamaguni-to-Kyoto operation as a whole. Fujita reveals in sensitive detail the many forms and levels of social tension within and between the several groups engaged in wood extraction and marketing. What gets lost in his examination (and in this body of literature as a whole) is the overall sense of lumbering as a system based on human interdependence and the yet broader sense of lumbering as a system of ecological exploitation in which a diverse group of humans, working together more or less awkwardly, manipulate a piece of their environment to their own mutual advantage. As a predictable corollary, one learns little from these works about the broader biosystemic consequences of this woodland manipulation. For the narrower industrial history focus of this study, however, these limitations present no problems.

2. The word *Yamaguni* is used in different ways by different scholars because local usages have changed over the years. The timber-growing watershed upstream from Shūzan is the area examined in Motoyoshi, *Senshin,* and Dōshisha daigaku jinbun kagaku kenkyūjo, comp., *Ringyō sonraku no shiteki kenkyū.* It should be borne in mind, however, that the "fifty-two-village lumbermen's organization" examined by Fujita, *Kinsei mokuzai,* included many other villages, mostly downstream between Shūzan and Tonoda.

3. Mizutani, "Kinsei Kyōto no mokuzai yusō (jō)," 45.

4. *Kabusugi* culture has two notable virtues: it permits cropping of steep hillsides with minimal disturbance of the soil and it allows rapid restocking of pole timber because the new growth is supported by preexisting root systems. After three or four cycles of regeneration, however, *sugi*

stumps lose their vitality, and seedlings must replace them. On the rise of tree planting (regenerative forestry or plantation silviculture) as a major source of new timber in Yamaguni, see, in particular, Motoyoshi, *Senshin,* 268–319. Also Fujita, *Kinsei mokuzai,* 115–123; Motoyoshi, "Kyōtofu Yamaguni chihō ni okeru rin'ya shoyū no keisei to zōrin no shiteki hatten (2)," 74–80; and idem, "Yamaguni ringyō chitai ni okeru jinkō zōrin no shinten to ikurin gijutsu no hensen," 78–83.

5. Motoyoshi, "Yamaguni ringyō chitai ni okeru jinkō zōrin no shinten to ikurin gijutsu no hensen," 95. Fujita, *Kinsei mokuzai,* 151–152, estimates that a Yamaguni raft consisted of 200–300 sticks. At 100 maritime *koku* per raft, during the nineteenth century the river carried some 80–100,000 *koku* (22,000–28,000 cubic meters) of Yamaguni timber annually.

6. Fujita, *Kinsei mokuzai,* 158. The official rafting season ran from 8/ 16 to 4/8, old calendar. Raftsmen managed to get the season lengthened slightly in the early nineteenth century, when raft volume was greater. Mizutani, "Kinsei Kyōto no mokuzai yusō (jō)," 49; "(ge)," 59–60.

7. Fujita, *Kinsei mokuzai,* 151–152, reports that if one adds rafts of bamboo and rafts from other sources above Hōtsu to Yamaguni timber, the number passing Hōtsu during the nineteenth century fluctuated erratically between about 1,700 and 2,300 per year. Generally this wood from other sources was of less value, much being *kuri, matsu,* and *zōki* (chestnut, pine, and "miscellaneous"). Fujita adds, however, that besides the records from Hōtsu, which give these figures, other rafts stopped at Yamamoto, so the total of Ōi rafts may have been about 3,000 per year. At 200–300 sticks per raft, this amounts to some 600,000–900,000 sticks per year. At 100 *koku* per raft, the Ōi freighted some 80,000 cubic meters of wood and bamboo annually through the gorges.

8. Mizutani, "Kinsei Kyōto no mokuzai yusō (jō)," 48. Motoyoshi, *Senshin,* 223–225. On the other hand, compared to other streams in Japan, the Ōi was not particularly bad as a rafting river. On the adequacy of rivers for log transport, in addition to the references on rafting cited in chapter 3, see the two articles by Matsumura, "Ōme no ringyō" and "Ōme ringyō ni okeru ikada," for descriptions of Tamagawa rafting.

9. This material on dams is from Fujita, *Kinsei mokuzai,* 155–164.

10. Ibid., 165–171, 250; Mizutani, "Kinsei Kyōto no mokuzai yusō (jō)," 44. Motoyoshi, "Yamaguni ringyō chitai ni okeru jinkō zōrin no shinten to ikurin gijutsu no hensen," 95, uses the 100 maritime *koku* figure. However, Nishikawa, "Ryūtsū ichiba kara mita mokuzai shōhin seisan no hatten," 7, speaks of 250 logs, amounting to 170–200 *koku* (perhaps meaning the dry grain *koku,* which is about 60 percent of the maritime *koku*) of lumber for rafts on the Hōtsu-Saga run. Mizutani, "Kinsei Kyōto no mokuzai yusō (ge)," 60, whose figures are sometimes perplexing, says that spring rafts generally consisted of 230 sticks apiece, and autumn rafts about 239.

11. Fujita, *Kinsei mokuzai,* 240.

12. Ibid., 248–250, 258–264, 320. From its establishment in about 1640 until 1664, the Utsune *unjōsho* evidently was under direct *bakufu* supervision; after 1664, Kameyama handled it.

13. Mizutani, "Kinsei Kyōto no mokuzai yusō (ge)," 59. Much of the Saga lumber canal exists today, still fed by the Hōtsugawa.

14. Ibid., 56–57.

15. Mizutani, "Kinsei Kyōto no mokuzai yusō (jō)," 54–55, "(ge)," 58–59.

16. Mizutani, "Kinsei Kyōto no mokuzai yusō (jō)," 54–55.

17. On timber scarcity and reforestation in Yamaguni, see Motoyoshi, *Senshin,* 269–273, 287.

18. Three reconstructions of this pre-1600 Yamaguni timber history are the pioneer work by Nishikawa, "Ringyō keizai shiron (2)"; the essay by Nakamura Ken, "Chūsei Yamagunishō no myō taisei"; and more recently Motoyoshi, *Senshin,* 13–60. In his book, Motoyoshi makes good use of the earlier Dōshisha work, which is a coordinated collection of essays that remains the best overall community study of the Yamaguni locality.

19. The term *yaku* (like the older term *shiki*) denotes a reciprocal relationship that embodies the two English concepts of "right" and "duty" and seems usefully translated as "function." The wood-cutting function was known as *soma yaku* or *yama yaku,* as well as *ono yaku.*

20. The nature and evolution of forest possession arrangements in Yamaguni are carefully examined in these five studies: Fujita, *Kinsei mokuzai,* 76–113; the essay by Oka Mitsuo, "Kinsei Yamaguni gō no ringyō keiei"; Motoyoshi, "Kyōto-fu Yamaguni chihō ni okeru rin'ya shoyū no keisei to zōrin no shiteki hatten (2)"; idem, "Yamaguni ringyō chitai ni okeru rin'ya shoyū no keisei to sono hensen"; and idem, "Kyoto-shi hokubu Ōigawa jōryū chiki ni okeru ringyō no tenkai."

21. Fujita, *Kinsei mokuzai,* 172–183.

22. Motoyoshi, *Senshin,* 233–238; Fujita, *Kinsei mokuzai,* 177–183. Fujita sees this organization as paralleling rather than supplanting the fifty-two-village *dantai.*

23. Motoyoshi, *Senshin,* 246–249, 273–275, 287–308. This change in the character of Yamaguni lumbermen—from a local elite whose position presumably was rooted in customary privilege to one rooted in mercantile strength—is a central focus of the works I have perused, especially the works of Motoyoshi. In his essay "Kyoto-shi hokubu Ōigawa jōryū chiki ni okeru ringyō no tenkai," Motoyoshi shows how this situation worked to the disadvantage of villagers in the uppermost reaches of the Ōi.

24. On *nenkiyama,* see Totman, *Green Archipelago,* 161–163.

25. Fujita, *Kinsei mokuzai,* 151–152, 164.

26. Motoyoshi, *Senshin,* 250–254.

27. Ibid., 251. Workmen valued the employment highly, but their pay was not exceptional: they earned some 2 to 3.5 *monme* per day, which was a fairly standard wage for laborers during the Edo period. In the final chaotic years of the Tokugawa regime their wages rose with inflation to about 9–15 *monme* per day by 1867, which kept their daily purchasing power at about the 3–4 *shō* of rice that it had been all along. Motoyoshi, *Senshin,* 254, 257.

28. Mizutani, "Kinsei Kyōto no mokuzai yusō (jō)," 48.

29. Fujita, *Kinsei mokuzai,* 159–161.

30. Motoyoshi, *Senshin,* 293–294; Fujita, *Kinsei mokuzai,* 262.

31. Thus, whereas the two-day trip from Tonoda to Utsune, handled by a three-man crew from Yamaguni, cost 9 *bu* per man-day (or 5 *monme,* 4 *bu* total), the one-day, three-man trip from Hōtsu to Saga cost 42 *bu* per man-day (or 12 *monme,* 6 *bu* total), almost five times as high an hourly wage and more than twice as much in total cost. Fujita, *Kinsei mokuzai,* 250–251.

32. Ibid., 250–251, 325–326. Fujita examines rafting disputes in detail. On 286 he estimates that a Hōtsu raft crewman earned about 1 to 1.4 *koku* per year from his river work, rather less than a princely sum.

33. Mizutani, "Kinsei Kyōto no mokuzai yusō (jō)," 47. His summary of the problems of Ōi shipping on 47–49 is lucid and concise.

34. Fujita, *Kinsei mokuzai,* 137–138, 227–230, 235–240.

35. Ibid., discusses these problems at 256–257, 265–270, 288, 294–307, 320–324, presenting evidence that the ebb and flow in Hōtsu-Yamamoto communal cohesiveness bore directly on the guild's effectiveness in negotiations with outsiders.

36. This material on the three-landings wholesalers is largely derived from Fujita, *Kinsei mokuzai,* 340–372.

37. Motoyoshi, *Senshin,* 226.

38. Ibid., 266.

39. Fujita, *Kinsei mokuzai,* 372.

40. Ibid., 355. Mizutani, "Kinsei Kyōto no mokuzai yusō (ge)," 54.

41. Mizutani, "Kinsei Kyōto no mokuzai yusō (jō)," 50; idem, "(ge)," 68.

42. Fujita, *Kinsei mokuzai,* 347–348. Tokoro, *Kinsei ringyōshi no kenkyū,* 293, defines *shiraki.* Besides these lumberyards, there were the purveyors of charcoal *(sumiya)* and firewood *(makiya),* who were important commercial figures in their own right.

43. Whether the city merchants purchased or brokered most of their goods by sealed bid, fixed price, or negotiated price is unclear.

44. Fujita, *Kinsei mokuzai,* 347–352.

45. Ibid., 370–372.

46. Motoyoshi, *Senshin,* 245, n. 3.

47. Fujita, *Kinsei mokuzai,* 390, 397–401; Mizutani, "Kinsei Kyōto no mokuzai yusō (jō)," 53–54. Mizutani notes that the Kyoto lumber situation was worsened by Osaka merchants who purchased Yamaguni timber, thereby pushing up the prices of what went to Kyoto.

Notes to Chapter 5

1. Whereas a synecological study attempts to view an ecosystem holistically as a functioning entity, an autecological study views it from the perspective of one of its particular species or biological communities.

2. See Totman, *Green Archipelago,* for fuller exploration of the social and environmental effects of deforestation.

3. This topic is discussed in Totman, *Origins.*

4. These topics, and those of the following few paragraphs, are developed in Totman, *Early Modern Japan,* especially chapters 12 and 13.

5. Nishikawa Zensuke developed this theme in his studies. The monographic study of Yoshino lumbering by Kasai Kyōetsu, *Yoshino ringyō no hatten kōzō,* was a major contribution to this perspective. In more general studies, Yoshida, *Kiba no rekishi,* 134, 150, characterizes advance payments as an expression of merchant capital controlling production, although he notes that some wholesalers opposed them to minimize competition among themselves. Funakoshi, *Nihon ringyō hattenshi,* 45, describes advance payments as devices that stopped producers from selling directly to brokers, and he speaks of the *myōga* ("thank you") payments made by merchants to government as serving the same sort of deterrent function.

6. The work of Izumi Eiji on Yoshino (see appendix C and the Bibliography) exemplifies this interpretation. Shimada, *Edo-Tōkyō,* 552, notes that in the later Edo period more wholesalers listed themselves as residents of rural areas, and that often they were people with other businesses as well. He also notes (573–574) a growing involvement of shippers in price setting and marketing in Edo, and reports (589) that the interest rate on advance payments fell from roughly 15 percent in 1811 to about 12 percent in 1820 and 1852.

7. See in particular Shimada, "Bakumatsu no goyōzai shidashibito Shinanoya Shōzaburō no gyōtai (I)."

8. Regarding the Kinugawa, see particularly Abe, "Kinsei goki Kita Kantō ni okeru ringyō no hatten." On Naguri, see Wakino, "Kinsei Nishikawa ringyō ni okeru zaimokushō keiei," and Katō, "Edo chimawari sanson no gōnō keiei." On the upper Ōi, see Motoyoshi, "Kyōto-shi hokubu Ōigawa jōryū chiiki ni okeru ringyō no tenkai," and on the Yoshino, Izumi Eiji, "Koseisansha to sonraku kyōdōtai."

9. On the use of packhorses, see Wigen, *Japanese Periphery.*

10. Andō, *Kinsei kōgaishi no kenkyū,* 78–81 and passim, notes the presence of this problem in various places, especially Kyoto.

11. Totman, *Green Archipelago,* 183.

12. Minami Ryoshin, "Waterwheels in the Preindustrial Economy of Japan," 1–15.

13. Andō, *Kinsei kōgaishi,* 303–333.

14. A solid source of information is Geological Survey of Japan, comp., *Geology and Mineral Resources of Japan.*

BIBLIOGRAPHICAL ESSAY

THE NOTES of this volume list many of the books and articles that examine the Tokugawa lumber industry. The notes to chapter 1 identify English-language works treating the broader economy. It may be helpful here to mention other bibliographical resources that will guide interested readers to further works in the two languages.

The Literature in English

The most valuable guide to English-language works on the early modern economy is William D. Wray, *Japan's Economy: A Bibliography of Its Past and Present* (NY: Markus Wiener, 1989), 7–35. The most valuable general bibliography on Tokugawa history is John W. Dower, *Japanese History and Culture from Ancient to Modern Times: Seven Basic Bibliographies*, 2d ed. (NY: Markus Wiener, 1994). A selective listing of books and articles published before 1960 is Bernard Silberman, *Japan and Korea: A Critical Bibliography* (Tucson: University of Arizona Press, 1962). Additional titles may be found in the notes and bibliographies of recent works, most comprehensively, Conrad Totman, *Early Modern Japan* (Berkeley: University of California Press, 1993). To stay abreast of current scholarship, one can consult the book reviews and articles contained in the newest issues of scholarly periodicals, notably the *Journal of Japanese Studies, Monumenta Nipponica, Acta Asiatica, Harvard Journal of Asiatic Studies,* and *Journal of Asian Studies.*

Readers planning a more thorough survey of the available literature will find the following works useful. The annual *Bibliography of Asian Studies* (Ann Arbor, MI: Association for Asian Studies) provides an exhaustive, topically ordered listing of books and articles published during the preceding year. The *Bibliography*'s preparation time is so great, however, that the most

137

recent issue covers works published about five years earlier. Moreover, users will find that as punditry and semischolarly commentary on Japan have proliferated in recent years, the *Bibliography*'s editors have been compelled to include more and more items of doubtful utility and to multiply their entry categories, with the result that reliable works dealing with early modern Japan have become scattered and more difficult to locate. Nevertheless, the *Bibliography* remains an indispensable listing of recent scholarship.

The association produced two cumulative bibliographies titled *Cumulative Bibliography of Asian Studies*, one in four author volumes and four subject volumes that covers books and articles published during the years 1941–1965 and one in three author and three subject volumes for the years 1966–1970. Other exhaustive bibliographical series cover the decades before 1941, but they list few usable materials on the Tokugawa economy. Interested readers will find these older bibliographies discussed in Silberman, *Japan and Korea*.

One reference work that merits note is the *Encyclopedia of Japan*, 7 vols. (Tokyo: Kodansha, 1983). It provides a few longer essays on broader aspects of early modern history and a substantial number of brief articles that contain information on specific topics, some of them economic in nature. Entries relating to early modern forest industries are minimal, however.

The Literature in Japanese

Needless to say, the primary sources on Japan's early modern economy are all in Japanese, save for a handful of materials relating to foreign trade. The secondary literature also is immense compared to what is available in English.

Two basic bibliographical guides provide the most complete and orderly access to scholarship and source materials on Tokugawa economic history. One is Honjō Eijirō, comp., *Nihon keizaishi dai 1* [2, 3 etc.] *bunken* [The literature of Japanese economic history, vol. 1ff] (7 vols. to date) (Kyoto: Nihon keizaishi kenkyūjo of Osaka keizai daigaku, 1933–). Each successive volume lists, in elaborate topical order, books, articles, and primary materials published since the preceding volume.

The other guide is Shigakkai, ed., *Shigaku zasshi* [Journal of historical studies] (Tokyo: Shigakkai of the Tokyo daigaku bungakubu). This journal, which is issued several times yearly, contains in its well-organized bibliographical sections detailed listings of recent books, articles, and newly published primary works on Japanese (and other) history. In addition, one issue each year is devoted to topically organized bibliographical essays in which specialists discuss scholarship of the preceding year in their field of expertise.

Other historical journals also provide bibliographical listings, but none is as complete as that of *Shigaku zasshi*. By methodical use of the two works above, a scholar can prepare an exhaustive listing of secondary materials and obtain guidance to primary sources on most topics in Tokugawa economic history.

These bibliographies will guide one to the works on early modern economic history in general and the lumber industry more specifically. A faster and topically more structured entrée to the pre-1980 scholarship on pre-Meiji forestry is the Bibliographical Essay in Conrad Totman, *The Green Archipelago: Forestry in Preindustrial Japan* (Berkeley: University of California Press, 1989) and the exhaustive listing of Japanese books and articles in the Bibliography.

Many Japanese scholarly journals publish essays on Tokugawa economic history, as a perusal of listings in *Shigaku zasshi* will reveal. On the history of lumbering more specifically, the preeminent journal during recent decades has been *TRK Kenkyū kiyō,* published by the Tokugawa rinseishi kenkyūjo in Tokyo. Most of the articles in its thick annual issues deal with topics in early modern forest history, although some are on modern forest history and others on related topics in early modern history.

BIBLIOGRAPHY

Abbreviations for Frequently Cited University Presses

CUP	(London): Cambridge University Press
	(New York): Columbia University Press
HUP	Harvard University Press
PUP	Princeton University Press
SUP	Stanford University Press
UCP	(Berkeley): University of California Press
	(Chicago): University of Chicago Press
UTP	University of Tokyo Press
YUP	Yale University Press

Abbreviations of Journals in English:

AA	Acta Asiatica
EEH	Explorations in Economic History
ESQ	Economic Studies Quarterly
JAS	Journal of Asian Studies
JEH	Journal of Economic History
JJS	Journal of Japanese Studies
KUER	Kyoto University Economic Review
MAS	Modern Asian Studies
MN	*Monumenta Nipponica*
OEP	Osaka Economic Papers
TASJ	Transactions of the Asiatic Society of Japan

Abbreviations of Journals in Japanese:

RK	*Ringyō keizai*
TRK Kenkyū kiyō	Tokugawa rinseishi kenkyūjo kenkyū kiyō

English-language Sources

AA 39 (1980). "Studies in the History of Japanese Currency Systems."

Akimoto, Hiroya. "Capital Formation and Economic Growth in Mid-nineteenth-Century Japan." *EEH* 18, 1 (Jan. 1981): 40–59.

Association for Asian Studies, comp. *Bibliography of Asian Studies.* Ann Arbor: AAS, annual.

————. *Cumulative Bibliography of Asian Studies, 1941–1965,* and *1966–1970.* Ann Arbor: AAS, 1969 and 1973.

Bellah, Robert N. *Tokugawa Religion.* Glencoe, IL: The Free Press, 1957.

Berry, Mary Elizabeth. *Hideyoshi.* Cambridge, MA: HUP, 1982.

Bix, Herbert P. *Peasant Protest in Japan, 1590–1884.* New Haven: YUP, 1986.

Borton, Hugh. "Peasant Uprisings in Japan of the Tokugawa Period." *TASJ,* 2d ser., 16 (1938): 1–219.

Bowen, Roger, ed., *E. H. Norman, His Life and Scholarship.* Toronto: University of Toronto Press, 1984.

Broadbridge, Seymour. "Economic and Social Trends in Tokugawa Japan." *MAS* 8, 3 (1974): 347–372.

Chambliss, William J. *Chiaraijima Village: Land Tenure, Taxation, and Local Trade, 1818–1884.* Association for Asian Studies Monograph, no.19. Tucson: University of Arizona Press, 1965.

Coaldrake, William H. "Edo Architecture and Tokugawa Law." *MN* 36, 3 (Autumn 1981): 235–284.

Coulborn, Rushton. *Feudalism in History.* Princeton: PUP, 1956.

Crawcour, E. S. "Changes in Japanese Commerce in the Tokugawa Period." *JAS* 22, 4 (Aug. 1963). Reprinted in Hall and Jansen, eds., *Studies,* 189–202.

————. "The Development of a Credit System in Seventeenth-Century Japan." *JEH* 21, 3 (Sep. 1961): 342–360.

————. "Kawamura Zuiken: A Seventeenth-Century Entrepreneur." *TASJ,* 3d ser., 9 (1966): 28–50.

————. "Notes on Shipping and Trade in Japan and the Ryukyus." *JAS* 23, 3 (May 1964): 377–381.

————. "The Premodern Economy." In *An Introduction to Japanese Civilization,* ed. Arthur Tiedemann, 461–486. NY: CUP, 1973.

————. "The Tokugawa Heritage." In *The State and Economic Enterprise,* ed. William W. Lockwood, 17–44. Princeton: PUP, 1965.

————. "The Tokugawa Period and Japan's Preparation for Modern Economic Growth." *JJS* 1, 1 (Autumn 1974): 113–125.

————, trans. "Some Observations on Merchants." *TASJ,* 3d ser., 8 (1961): 1–139.

Daniels, Gordon, ed. *Europe Interprets Japan.* Tenterden, Kent: Paul Norbury, 1984.

Dore, Ronald P. *Education in Tokugawa Japan.* Berkeley: UCP, 1965.

Dower, John W., and Timothy S. George. *Japanese History and Culture from Ancient to Modern Times: Seven Basic Bibliographies.* 2d ed. NY: Markus Wiener, 1994.

———, ed. *Origins of the Modern Japanese State: Selected Writings of E. H. Norman.* NY: Pantheon, 1975.

Duus, Peter. *Feudalism in Japan.* NY: Alfred A. Knopf, 1969.

Encyclopedia of Japan. 7 vols. Tokyo: Kodansha, 1983.

Eyre, John D. "Patterns of Japanese Salt Production and Trade." *Occasional Papers,* no. 3, 15–46. Ann Arbor: University of Michigan Center for Japanese Studies, 1952.

Flershem, Robert G. "The Indomitables: Fishermen Outcastes of Tokugawa Japan." In *Proceedings of the Sixth Symposium on Asian Studies,* 657–676. Hong Kong: Asian Research Service, 1984.

———. "Some Aspects of Japan Sea Shipping and Trade in the Tokugawa Period, 1603–1867." *Proceedings of the American Philosophical Society* 110, 3 (1966): 182–226.

———. "Some Aspects of Japan Sea Trade in the Tokugawa Period." *JAS* 23, 3 (May 1964): 405–416.

Fruin, W. Mark. "The Firm as Family and the Family as Firm in Japan." *Journal of Family History* 5, 4 (Winter 1980): 432–449.

Furushima, Toshio. "The Village and Agriculture during the Edo Period." In Hall, ed., *Cambridge History,* 478–518.

Geological Survey of Japan, comp. *Geology and Mineral Resources of Japan.* 2d ed. Kawasaki: Geological Survey of Japan, 1960.

Gluck, Carol. "The Past in the Present." In *Postwar Japan as History,* ed. Andrew Gordon, 64–95. Berkeley: UCP, 1993.

Hall, John Whitney. "The Castle Town and Japan's Modern Urbanization." In Hall and Jansen, eds., *Studies,* 169–188. Reprinted from *Far Eastern Quarterly* 15, 1 (Nov. 1955).

———. "Changing Conceptions of the Modernization of Japan." In *Changing Japanese Attitudes toward Modernization,* ed. Marius B. Jansen, 7–41. Princeton: PUP, 1965.

———. "Feudalism in Japan—A Reassessment." *Comparative Studies in Society and History* 5, 1 (Oct. 1962): 15–51.

———. "The Nature of Traditional Society: Japan." In *Political Modernization in Japan and Turkey,* ed. Robert E. Ward and Dankwort A. Rustow, 14–41. Princeton: PUP, 1964.

———, ed. *The Cambridge History of Japan.* Vol. 4, *Early Modern Japan.* Cambridge: CUP, 1991.

Hall, John W., and Marius B. Jansen, eds. *Studies in the Institutional History of Early Modern Japan.* Princeton: PUP, 1968.

Hanley, Susan B. "A High Standard of Living in Nineteenth-Century Japan: Fact or Fantasy?" *JEH* 43, 1 (1983): 183–192.

———. "How Well Did the Japanese Live in the Tokugawa Period? A Historian's Reappraisal." *ESQ* 38, 4 (Dec. 1987): 309–322.

———. "Tokugawa Society: Material Culture, Standard of Living, and Life-styles." In Hall, ed., *Cambridge History,* 660–705.

———. "Urban Sanitation in Preindustrial Japan." *Journal of Interdisciplinary History* 18, 1 (Summer 1987): 1–26.

Hanley, Susan B., and Arthur P. Wolf, eds. *Family and Population in East Asian History.* Stanford: SUP, 1985.

Hanley, Susan B., and Kozo Yamamura. *Economic and Demographic Change in Preindustrial Japan, 1600–1868.* Princeton: PUP, 1977.

———. "A Quiet Transformation in Tokugawa Economic History." *JAS* 30, 2 (Feb. 1971): 373–384.

Harootunian, H. D. "America's Japan/Japan's Japan." In Miyoshi and Harootunian, eds., *Japan in the World*, 196–221.

Hauser, William B. "The Diffusion of Cotton Processing and Trade in the Kinai Region in Tokugawa Japan." *JAS* 33, 4 (Aug. 1974): 633–649.

———. *Economic Institutional Change in Tokugawa Japan: Osaka and the Kinai Cotton Trade.* Cambridge: CUP, 1974.

———. "Osaka: A Commercial City in Tokugawa Japan." *Urbanism Past and Present* 5 (Winter 1977–1978): 23–36.

———. "Osaka Castle and Tokugawa Authority in Western Japan." In *The Bakufu in Japanese History*, ed. Jeffrey P. Mass and Hauser, 153–172. Stanford: SUP, 1985.

Hayami, Akira, ed. *Preconditions to Industrialization in Japan.* Papers from Ninth International Economic History Congress, Bern, 1986.

Honjō, Eijirō. *Economic Theory and History of Japan in the Tokugawa Period.* Tokyo: Maruzen, 1943. Republished NY: Russell and Russell, 1965.

———. *The Social and Economic History of Japan.* Kyoto: Institute for Research in the Economic History of Japan, 1935. Republished NY: Russell and Russell, 1965.

Horie, Yasuzo. "The Feudal States and the Commercial Society in the Tokugawa Period." *KUER* 28, 2 (Oct. 1958): 1–16.

Hoston, Germaine. *Marxism and the Crisis of Development in Prewar Japan.* Princeton: PUP, 1986.

Howell, David L. "Hard Times in the Kantō: Economic Change and Village Life in Late Tokugawa Japan." *MAS* 23, 2 (May 1989): 349–371.

———. *Indigenous Capitalism in Nineteenth-Century Japan: The Transformation of the Hokkaido Fishery.* Berkeley: UCP, 1994.

———. "Proto-Industrial Origins of Japanese Capitalism." *JAS* 51, 2 (May 1992): 269–286.

Important Trees of Japan. Natural Resources Section, report no. 119. Tokyo: General Headquarters, Supreme Commander for the Allied Powers, 1949.

Innes, Robert Leroy. *The Door Ajar: Japan's Foreign Trade in the Seventeenth Century.* 2 vols. Ann Arbor: University Microfilms, 1980.

Irie, Hiroshi. "Apprenticeship Training in Tokugawa Japan." *AA* 54 (1988): 1–23.

Jannetta, Ann Bowman. *Epidemics and Mortality in Early Modern Japan.* Princeton: PUP, 1987.

Johnson, Linda L. "Patronage and Privilege: The Politics of Provincial Capitalism in Early Modern Japan." Ph.D. dissertation, Stanford University, 1983.

———. "Prosperity and Welfare: The Homma Family and Agricultural Improvement in Meiji Japan." *TASJ*, 4th ser., 5 (1990): 1–24.

Kada Yukiko. "The Evolution of Joint Fisheries Rights and Village Community Structure on Lake Biwa, Japan." In Ruddle and Akimichi, eds., *Maritime Institutions*, 137–158.

Kalland, Arne. "A Credit Institution in Tokugawa Period: The *Ura-Tamegin* Fund of Chikuzen Province." In Daniels, ed., *Europe Interprets Japan*, 3–12.

———. *Fishing Villages in Tokugawa Japan*. Honolulu: University of Hawai'i Press, 1994.

———. "Pre-modern Whaling in Northern Kyushu." In Pauer, ed., *Silkworms, Oil, and Chips*, 8:29–50.

———. "Sea Tenure in Tokugawa Japan: The Case of Fukuoka Domain." In Ruddle and Akimichi, eds., *Maritime Institutions*, 11–36.

Katsu, Kokichi. *Musui's Story.* Tucson: University of Arizona Press, 1991.

Kelly, William W. *Deference and Defiance in Nineteenth-Century Japan*. Princeton: PUP, 1985.

———. *Water Control in Tokugawa Japan: Irrigation Organization in a Japanese River Basin, 1600–1870*. East Asia Papers, no. 31. Ithaca: Cornell University China-Japan Program, 1982.

Koschmann, J. Victor. *The Mito Ideology.* Berkeley: UCP, 1987.

Leupp, Gary. *Servants, Shophands, and Laborers in the Cities of Tokugawa Japan*. Princeton: PUP, 1992.

Marshall, Byron. *Nationalism and Capitalism in Prewar Japan*. Stanford: SUP, 1967.

McClain, James L. "Castle Towns and Daimyo Authority: Kanazawa in the Years 1583–1630." *JJS* 6, 2 (Summer 1980): 267–299.

———. *Kanazawa*. New Haven: YUP, 1982.

McEwan, J. R. *The Political Writings of Ogyū Sorai*. Cambridge: CUP, 1962.

Metzger-Court, Sarah. "Towards National Integration: A Comparative Survey of Economic Progress in the Prefectures of Wakayama, Okayama and Hiroshima During the Nineteenth Century." In Daniels, ed., *Europe Interprets Japan*, 13–19.

Minami, Ryoshin. "Waterwheels in the Preindustrial Economy of Japan." *Hitotsubashi Journal of Economics* 22, 2 (Feb. 1982): 1–15.

Miyamoto, Mataji. "Economic and Social Development of Osaka." *OEP* 3, 1 (Dec. 1954): 11–28.

———. "The Merchants of Osaka." *OEP* 7, 1 (Sep. 1958): 1–13.

———. "The Merchants of Osaka: II." *OEP* 15, 28 (Nov. 1966): 11–32.

Miyoshi, Masao, and H. D. Harootunian, eds. *Japan in the World*. Durham: Duke University Press, 1993.

Mori, Yasuhiro. "Loans to Daimyos by the Osaka Money Changers." *OEP* 15, 29 (Mar. 1967): 19–28.

Moriya, Katsuhisa. "Urban Networks and Information Networks." In Nakane and Ōishi, eds., *Tokugawa Japan*, 97–123.

Morris-Suzuki, Tessa. "Concepts of Nature and Technology in Pre-Industrial Japan." *East Asian History,* no. 1 (June 1991): 81–97.

————. "Sericulture and the Origins of Japanese Industrialization." *Technology and Culture* 33, 1 (Jan. 1992): 101–121.

Nagahara, Keiji, and Kozo Yamamura. "Shaping the Process of Unification: Technological Progress in Sixteenth- and Seventeenth-Century Japan." *JJS* 14, 1 (Winter 1988): 77–109.

Najita, Tetsuo. "Japan's Industrial Revolution in Historical Perspective." In Miyoshi and Harootunian, eds., *Japan in the World,* 13–30.

————. "Political Economism in the Thought of Dazai Shundai (1680–1747)." *JAS* 31, 4 (Aug. 1972): 821–839.

————. *Visions of Virtue in Tokugawa Japan.* Chicago: UCP, 1987.

Nakai, Nobuhiko. "Commercial Change and Urban Growth in Early Modern Japan." In Hall, ed., *Cambridge History,* 519–595.

Nakamura, J. I. "Human Capital Accumulation in Premodern Rural Japan." *JEH* 41, 2 (June 1981): 263–281.

Nakamura, Satoru. "The Development of Rural Industry." In Nakane and Ōishi, eds., *Tokugawa Japan,* 81–96.

Nakane, Chie, and Shinzaburō Ōishi, eds., *Tokugawa Japan.* Tokyo: UTP, 1990.

Nishikawa, Shunsaku. "The Economy of Chōshū on the Eve of Industrialization." *ESQ* 38, 4 (Dec. 1987): 323–337.

————. "Productivity, Subsistence, and By-employment in the Mid-Nineteenth-Century Chōshū." *EEH* 15 (1978): 69–83.

————. "Protoindustrialization in the Domain of Chōshū in the Eighteenth and Nineteenth Centuries." *Keio Economic Studies* 18, 2 (1981): 13–26.

Norman, E. H. *Japan's Emergence as a Modern State.* NY: Institute of Pacific Relations, 1940.

Pauer, Erich, ed. *Silkworms, Oil, and Chips: Proceedings of the Economics and Economic History Section of the Fourth International Conference on Japanese Studies.* Japan Seminar, University of Bonn, 1986.

Pratt, Edward. "Village Elites in Tokugawa Japan: The Economic Foundations of the Gōnō." Ph.D. dissertation, University of Virginia, 1991.

Ramseyer, J. Mark. "Thrift and Diligence: House Codes of Tokugawa Merchant Families." *MN* 34, 2 (Summer 1979): 209–230.

Ravina, Mark J. "Political Economy and Statecraft in Early Modern Japan." Ph.D. dissertation, Stanford University, 1991.

Roberts, Luke S. "The Merchant Origins of National Prosperity Thought in Eighteenth-Century Tosa." Ph.D. dissertation, Princeton University, 1991.

Rozman, Gilbert. *Urban Networks in Ch'ing China and Tokugawa Japan.* Princeton: PUP, 1974.

Rubinger, Richard. *Private Academies of Tokugawa Japan.* Princeton: PUP, 1982.

Ruddle, Kenneth, and Akimichi Tomoya, eds. *Maritime Institutions in the Western Pacific.* Senri Ethnological Studies, no. 17. Osaka: National Museum of Ethnology, 1984.

Sadler, Arthur L. *The Maker of Modern Japan: The Life of Tokugawa Ieyasu.* London: Allen and Unwin, 1937.

Saitō, Osamu. "Changing Structure of Urban Employment and Its Effects on Migration Patterns in Eighteenth- and Nineteenth-Century Japan." In Hayami, ed., *Pre-Conditions to Industrialization in Japan.*

———. "The Labor Market in Tokugawa Japan: Wage Differentials and the Real Wage Level, 1727–1830," *EEH* 15 (Jan. 1978): 84–100.

———. "Population and the Peasant Economy in Proto-Industrial Japan." *Journal of Family History* 8, 1 (Spring 1983): 30–54.

Sakudō, Yōtarō. "The Management Practices of Family Business." In Nakane and Ōishi, eds., *Tokugawa Japan,* 147–166.

Sansom, G. B. *Japan, A Short Cultural History.* London: Cresset Press, 1931.

Satō, Tsuneo. "Tokugawa Villages and Agriculture." In Nakane and Ōishi, eds., *Tokugawa Japan,* 37–80.

Sharpe, Grant W., Clare W. Hendee, and Shirley W. Allen. *Introduction to Forestry.* 4th ed. NY: McGraw-Hill, 1976.

Sheldon, Charles David. "'Pre-Modern' Merchants and Modernization in Japan." *MAS* 5, 3 (1971): 193–206.

———. *The Rise of the Merchant Class in Tokugawa Japan, 1600–1868.* Locust Valley, NY: J. J. Augustin, 1958.

Shimbo, Hiroshi. "A Study of the Growth of Cotton Production for the Market in the Tokugawa Era." *Kobe University Economic Review* 1 (1955): 55–69.

Silberman, Bernard. *Japan and Korea: A Critical Bibliography.* Tucson: University of Arizona Press, 1962.

Smith, Neil Skene. "An Introduction to Some Japanese Economic Writings of the Eighteenth Century." *TASJ* 2d ser., 11 (1934): 32–105.

Smith, Robert J. "Aspects of Mobility in Pre-Industrial Japanese Cities." *Comparative Studies in Society and History* 5 (July 1963): 416–423.

———. "Small Families, Small Households, and Residential Instability: Town and City in Pre-Modern Japan." In *Household and Family in Past Time,* ed. Peter Laslett, 429–471. London: CUP, 1972.

Smith, Thomas C. *The Agrarian Origins of Modern Japan.* Stanford: SUP, 1959.

———. "Farm Family By-Employments in Preindustrial Japan." In T. C. Smith, *Native Sources,* 71–102.

———. *Native Sources of Japanese Industrialization, 1750–1920.* Berkeley: UCP, 1988.

———. "Ōkura Nagatsune and the Technologists." In T. C. Smith, *Native Sources,* 173–198.

Takatera, Sadao, and Noboru Nishikawa. "Genesis of Divisional Management and Accounting Systems in the House of Mitsui, 1710–1730." *Accounting Historian's Journal* 11, 1 (Spring 1984): 141–149.

Tanaka, Stefan. *Japan's Orient: Rendering Pasts into History.* Berkeley: UCP, 1993.

Thunberg, Carl Peter. *Flora Japonica* (in Latin). Leipzig: I. G. Müller, 1784; NY: Oriole Editions, 1975.

Toby, Ronald P. "Both a Borrower and a Lender Be: From Village

Moneylender to Rural Banker in the Tempō Era." *MN* 46, 4 (Winter 1991): 483–512.

Tokoro, Mitsuo. *The Wood-Cutting and Transporting System of Kiso-Style.* Tokyo: Tokugawa Institute for the History of Forestry, 1977.

Totman, Conrad. *Early Modern Japan.* Berkeley: UCP, 1993.

———. *The Green Archipelago: Forestry in Preindustrial Japan.* Berkeley: UCP, 1989.

———. *The Origins of Japan's Modern Forests: The Case of Akita.* Honolulu: University of Hawai'i Press, 1985.

———. "Preindustrial River Conservancy [in Japan]: Causes and Consequences." *MN* 47, 1 (Spring 1992): 59–76.

———. *Tokugawa Ieyasu, Shogun.* South San Francisco: Heian International, 1983.

———. "Tokugawa Peasants: Win, Lose, or Draw?" *MN* 41, 4 (Winter 1986): 457–476.

Toyoda, Takeshi. *A History of Pre-Meiji Commerce in Japan.* Tokyo: Kokusai Bunka Shinkokai, 1969.

———. "Japanese Guilds." *Annals of the Hitotsubashi Academy 5,* 1 (Oct. 1954): 72–85.

Tsuji, Tatsuya. "Politics in the Eighteenth Century." In Hall, ed. *Cambridge History,* 425–477.

Tsukahira, Toshio G. *Feudal Control in Tokugawa Japan.* Cambridge, MA: HUP, 1966.

Vaporis, Constantine N. "Overland Communications in *Tokugawa Japan.*" Ph.D. dissertation, Princeton University, 1987.

———. "Post Station and Assisting Villages: Corvée Labor and Peasant Contention." *MN* 41, 4 (Winter 1986): 377–414.

Vlastos, Stephen. *Peasant Protests and Uprisings in Tokugawa Japan.* Berkeley: UCP, 1986.

Wakita, Osamu. "The Social and Economic Consequences of Unification." In Hall, ed., *Cambridge History,* 96–127.

Walthall, Anne. "Peripheries: Rural Culture in Tokugawa Japan," *MN* 39, 4 (Winter 1984): 371–392.

———. "Village Networks: Sōdai and the Sale of Edo Nightsoil." *MN* 43, 3 (Autumn 1988): 293–302.

Wigen, Kären. "The Geographic Imagination in Early Modern Japanese History: Retrospect and Prospect." *JAS* 51, 1 (Feb. 1992): 3–29.

———. *The Making of a Japanese Periphery, 1750–1920.* Berkeley: UCP, 1994.

Wileman, A. E. "Salt Manufacture in Japan." *TASJ* 1st ser., no. 17 (1888–1889): 1–66.

Wray, William D. *Japan's Economy: A Bibliography of Its Past and Present.* NY: Markus Wiener, 1989.

Yamakawa, Kikue. *Women of the Mito Domain.* Tokyo: UTP, 1992.

Yamamura, Kozo. "The Increasing Poverty of the Samurai in Tokugawa Japan, 1600–1868." *JEH* 31 (June 1971): 378–406.

———. "Returns on Unification: Economic Growth in Japan, 1550–

1650." In John W. Hall et al., eds., *Japan before Tokugawa,* 327–372. Princeton: PUP, 1981.

————. "Toward a Reexamination of the Economic History of Tokugawa Japan, 1600–1867." *JEH* 33, 3 (Sep. 1973): 509–546.

Yamamura, Kozo, and Tetsuo Kamiki. "Silver Mines and Sung Coins— A Monetary History of Medieval and [Early] Modern Japan in International Perspective." In *Precious Metals in the Later Medieval and Early Modern Worlds,* ed. J. F. Richards, 329–362. Durham: Carolina Academic Press, 1983.

Yasuba, Yasukichi. "The Tokugawa Legacy: A Survey." *ESQ* 38, 4 (Dec. 1987): 290–308.

Yazaki, Takeo. *Social Change and the City in Japan.* Tokyo: Japan Publications, 1968.

Japanese-Language Sources

Abe Akira. "Kinsei goki Kita Kantō ni okeru ringyō no hatten" [The development of the North Kantō lumber industry in the later Edo period]. In *Kinsei no shihai taisei to shakai kōzō* [The early modern control system and social structure], ed. Kitajima Masamoto, 511–548. Tokyo: Yoshikawa kōbunkan, 1983.

————. "Kinsei ni okeru Kinugawa jōryūiki no ikada nagashi" [Early modern lumber rafting on the upper Kinugawa]. *TRK Kenkyū kiyō* 60 (Mar. 1986): 271–309.

Andō Seiichi. *Kinsei kōgaishi no kenkyū* [A study of pollution problems in early modern Japan]. Tokyo: Yoshikawa kōbunkan, 1992.

Chiba Tokuji. *Hageyama no kenkyū* [A study of bald mountains]. Tokyo: Nōrin kyōkai, 1956.

Daijinmei jiten [Biographical dictionary]. Tokyo: Heibonsha, 1958.

Dōshisha daigaku jinbun kagaku kenkyūjo, comp. *Ringyō sonraku no shiteki kenkyū: Tanba Yamaguni ni okeru* [Historical studies of forest villages in Yamaguni district of Tanba prefecture]. Kyoto: Mineruva shoten, 1967.

Fujita Yoshihisa. "Yoshinogawa jōryūiki ni okeru kinsei no sonraku kōzō no seikaku to ikurin to hatten" [The character of villages in the upper Yoshino river basin and the rise of plantation forestry]. *TRK Kenkyū kiyō* 60 (Mar. 1986): 221–270.

Fujita Yoshitami. *Kinsei mokuzai ryūtsūshi no kenkyū: Tanbazai ryūtsū no hatten katei* [A study of the history of Edo-period lumber provisioning: The development of Tanba lumber distribution]. Tokyo: Ōhara shinseisha, 1973.

Funakoshi Shōji. *Nihon ringyō hattenshi* [A history of the development of the Japanese lumber industry]. Tokyo: Chikyū shuppansha, 1960.

Hirao Michio. *Tosa han ringyō keizaishi* [An economic history of Tosa forestry]. Kōchi: Kōchi shimin toshokan, 1956.

Honjō Eijirō, comp. *Nihon keizaishi dai 1* [2, 3, etc.] *bunken* [The literature of Japanese economic history, vol. 1ff]. 7 vols. to date. Kyoto: Nihon keizaishi kenkyūjo of Osaka keizai daigaku, 1933–.

Iioka Masatake. "Enshū Funagira ni okeru bakufu no kureki shobun" [Handling government split pieces at Funagira in Tōtōmi]. *TRK Kenkyū kiyō* 50 (Mar. 1976): 101–119.

———. "Enshū Funagira ni okeru bakufu yōzai no chūkei kinō" [Operation of the government tax-lumber transfer station at Funagira]. *TRK Kenkyū kiyō* 52 (Mar. 1978): 95–117.

———. "Kinsei chūki no yōzai seisan shihō to saiunhi" [Logging procedure and shipping costs of mid-Edo-period government lumber]. *TRK Kenkyū kiyō* 51 (Mar. 1977): 107–124.

———. "Takatōryō 'ohayashi' ni okeru goyōki-uriki no saishutsu: Genroku-Kyōhō ki no 'mokushi sato' " [The taking out of government and market lumber from government forests in Takatō domain: So-called logging villages circa 1700–1730]. *TRK Kenkyū kiyō* 47 (Mar. 1973): 447–463.

———. "Rinzai hokyūsaku yori mita bakufu no shukueki keiei [The *bakufu*'s handling of post stations as seen from the perspective of timber provisioning]. *TRK Kenkyū kiyō* 61 (Mar. 1987): 111–135.

Iwanaga Yutaka. "Edo-Meiji ki ni okeru Yoshino ringyō no ikurin gijutsu" [The afforestation technology of Yoshino forests in the Edo-Meiji era]. *RK* 255 (Jan. 1970): 14–23.

Iwasaki Naoto. *Akita ken Noshirogawa kami chihō ni okeru sugibayashi no seiritsu narabi ni kōshin ni kan suru kenkyū* [Research on the establishment and revitalization of cryptomeria forests in the upper Noshiro River area of Akita prefecture]. Tokyo: Kōrinkai, 1939.

Izumi Eiji. "Kinsei shoki ni okeru kenchiku yōshiki no tenkan to Yoshino ringyō" [Changes in architectural style and Yoshino forestry in the early Edo period]. *TRK Kenkyū kiyō*, vol. 23 (Mar. 1989): 27–48.

———. "Kinsei Yoshino chihō ni okeru yamamoto zaimoku shōnin no katsudō" [The activities of local lumber merchants in the early modern Yoshino region]. *TRK Kenkyū kiyō* 61 (Mar. 1987): 137–160.

———. "Koseisansha to sonraku kyōdōtai" [Small-scale producers and the village community]. *TRK Kenkyū kiyō*, vol. 24 (Mar. 1990): 89–120.

Kasai Kyōetsu. *Yoshino ringyō no hatten kōzō* [The development of forestry in Yoshino]. Published as *Gakujutsu hōkoku tokushū* 15. Utsunomiya: Utsunomiya daigaku nōgakubu, 1962.

Katō Morihiro. "Edo chimawari sanson no gōnō keiei" [The managerial practices of a rich mountain villager near Edo]. *TRK Kenkyū kiyō* 61 (Mar. 1987): 161–206.

———. "*Yoshino ringyō zensho* no kenkyū" [A study of the *Compendium on Yoshino Forestry*]. *TRK Kenkyū kiyō* 58 (Mar. 1984): 189–223.

———. "Nishikawa ringyō hasseishi ni kan suru ichi kōsatsu" [An examination of the beginnings of forestry in the Nishikawa area]. *TRK Kenkyū kiyō* 56 (Mar. 1982): 165–196.

Kitamura Shirō and Okamoto Shōgo. *Genshoku Nihon jumoku zukan* [Illustrated handbook of Japanese trees and shrubs]. Osaka: Hoikusha, 1959.

Matsumura Yasukazu. "Kinsei Ōme ringyō no seiritsu oyobi hatten ni kan suru rekishi chirigakuteki kenkyū (shōroku)" [A historico-geographical study of the establishment and evolution of forestry in Ōme during the Edo period: A summary]. *Tōkyō geijitsu daigaku kenkyū hōkoku* 16 (1964): 1–22.

————. "Ōme no ringyō" [The forest industry of the Ōme vicinity]. In *Nihon sangyōshi taikei,* ed. Chihōshi kenkyū kyōgikai, 4:183–199. Tokyo: Tōkyō Daigaku Shuppankai, 1959 and 1970.

————. "Ōme ringyō ni okeru ikada" [Rafting in Ōme lumbering]. *Jinbun chiri* 7, 5 (1955): 14–33.

Michishige Tetsuo. "Hansei goki ni okeru hanyō mokuzai no seisan kōzō" [The structure of lumber production for government use in the late Edo period]. *Geibi chihōshi kenkyū* 78 (1969): 1–10.

Mitsuhashi Tokio. "Yoshino-Kumano no ringyō" [Forestry in Yoshino and Kumano]. In *Nihon sangyōshi taikei,* ed. Chihōshi kenkyū kyōgikai, 6:241–266. Tokyo: Tokyo daigaku shuppankai, 1960.

Mizutani Seizō. "Kinsei Kyōto no mokuzai yusō (jō)—Umezu o chūshin to shita sankasho nakama no rekishi chiriteki kōsatsu" [The movement of lumber in Edo-period Kyoto: A historio-geographical examination of the monopoly merchants of the 'three places', especially Umezu: part 1]. *Ritsumeikan bungaku* 181 (1960): 43–61.

————. "Kinsei Kyōto no mokuzai yusō (ge)—Umezu o chūshin to shita sankasho nakama no rekishi chiriteki kōsatsu" [The movement of lumber in Edo-period Kyoto: A historio-geographical examination of the monopoly merchants of the "three places," especially Umezu: part 2]. *Ritsumeikan bungaku* 183 (1961): 53–77.

Motoyoshi Rurio. "Kyōtofu Yamaguni chihō ni okeru rin'ya shoyū no keisei to zōrin no shiteki hatten (2)" [The character of woodland possession rights and the historical development of afforestation in the Yamaguni area of Kyoto Prefecture, part 2]. *Kyōto furitsu daigaku nōgakubu enshūrin hōkoku* 13 (1969): 39–81.

————. "Kyōto-shi hokubu Ōigawa jōryū chiiki ni okeru ringyō no tenkai" [The development of lumbering in the upper Ōi River valley north of Kyoto city]. *TRK Kenkyū kiyō* 57 (Mar. 1983): 51–79.

————. *Senshin ringyō chitai no shiteki kenkyū: Yamaguni ringyō no hatten katei* [A historical study of an advanced area of forest industry: The development of the Yamaguni forest industry]. Tokyo: Tamagawa daigaku shuppanbu, 1983.

————. "Yamaguni ringyō chitai ni okeru jinkō zōrin no shinten to ikurin gijutsu no hensen" [The development of afforestation and changes in silviculture in the Yamaguni forest area]. *TRK Kenkyū kiyō* 55 (Mar. 1981): 72–97.

————. "Yamaguni ringyō chitai ni okeru rin'ya shoyū no keisei to sono hensen" [The character of, and changes in, woodland possession in the Yamaguni forest area]. *TRK Kenkyū kiyō* 56 (Mar. 1982): 95–119.

Nakamura Ken. "Chūsei Yamagunishō no myō taisei" [The titleholder system of the medieval Yamaguni estate]. In Dōshisha, comp., *Ringyō sonraku,* 29–82.

Nihon Gakushiin. *Meiji zen Nihon ringyō gijutsu hattatsushi* [The development of forest technology in pre-Meiji Japan]. Tokyo: Nōkan kagaku igaku kenkyū shiryōkan, 1980.

Nishikawa Zensuke. "Edo zaimokushō no kigen: Edo mokuzai ichiba-shi josetsu" [The origins of Edo lumber markets: An introduction to the history of the Edo lumber marketplace]. *RK* 169 (Nov. 1962): 4–17.

———. "Ringyō keizai shiron (1)" [A treatise on the economics of forestry: Part 1]. *RK* 134 (1959): 4–13.

———. "Ringyō keizai shiron (2)" [A treatise on the economics of forestry: Part 2]. *RK* 135 (1959): 15–30.

———. "Ryūtsū ichiba kara mita mokuzai shōhin seisan no hatten" [The development of commercial lumbering as seen from the marketplace]. *Shakai keizai shigaku* 27, 1 (1961): 1–24.

Oka Mitsuo. "Kinsei Yamaguni gō no ringyō keiei" [The administration of forest production in early modern Yamaguni district]. In Dōshisha, comp., *Ringyō sonraku*, 287–398.

Ōsaki Rokurō. "Ohayashi chiseki kakutei katei no ichi kenshō—Kōzuke-kuni (Gunma-ken) Tone-gun o taishō to shite" [On standardizing land surveys of *ohayashi*: With Tone district of Kōzuke as the subject]. *RK* 119 (1958): 19–32.

Ōta Katsuya. "Kinsei chūki no Ōsaka zaimoku ichiba—torihiki kiyaku to 'ton'ya' no nakama gitei no kentō o chūshin ni" [The mid-Edo period Osaka lumber market: An examination of the trading agreements and guild arrangements of wholesale merchants]. *TRK Kenkyū kiyō* 50 (Mar. 1976): 69–100.

Shigakkai, ed. *Shigaku zasshi* [Journal of historical studies]. Tokyo: Shigakkai of the Tokyo Daigaku Bungakubu.

Shimada Kinzō. "Bakuhan kenryoku kōzōka no zaimoku ton'ya nakama no kōdō" [The operation of lumber merchant guilds under the Tokugawa system]. *TRK Kenkyū kiyō* 61 (Mar. 1987): 1–52.

———. "Bakuhan taiseika no Edo zaimokushō no shōtai" [The business setup of Edo lumbermen in the Tokugawa system]. *TRK Kenkyū kiyō*, vol. 24 (Mar. 1990): 160–162.

———. "Bakuhan taiseika no Edo zaimokushō no shōtai II" [The business setup of Edo lumbermen in the Tokugawa system, part II]. *TRK Kenkyū kiyō*, vol. 26 (Mar. 1992): 79–112.

———. "Bakumatsu no goyōzai shidashibito Shinanoya Shōzaburō no gyōtai (I)" [The business situation of the late-Tokugawa official lumber purveyor Shinanoya Shōzaburō, part I]. *TRK Kenkyū kiyō* 58 (Mar. 1984): 59–82.

———. "Bakumatsu no goyōzai shidashibito Shinanoya Shōzaburō no gyōtai (II)" [The business situation of the late-Tokugawa official lumber purveyor Shinanoya Shōzaburō, part II]. *TRK Kenkyū kiyō* 59 (Mar. 1985): 35–87.

———. "Bakumatsu no goyōzai shidashibito Shinanoya Shōzaburō no gyōtai (III)" [The business situation of the late-Tokugawa official lumber purveyor Shinanoya Shōzaburō, part III]. *TRK Kenkyū kiyō* 60 (Mar. 1986): 97–142.

———. "Edo zaimoku ton'ya no funsō to futan kuyaku" [Disputes among Edo lumber merchants, and their public duties]. *TRK Kenkyū kiyō*, vol. 23 (Mar. 1989): 179–218.

———. *Edo-Tōkyō zaimoku ton'ya kumiai seishi* [An authoritative history of the lumberman's association of Edo-Tokyo]. Tokyo: Doi Ringaku Shinkōkai, 1976.

———. "Kawabe ichibangumi koton'ya kumiai monjo to Edo zaimoku ichiba" [The records of the original first riverside wholesalers' guild and the Edo lumber market]. *TRK Kenkyū kiyō* 51 (Mar. 1977): 37–62.

Shioya Junji. "Fushimi chikujō to Akita sugi" [Construction of Fushimi Castle and the cryptomeria of Akita]. In *Kokushi danwakai zasshi* (of Tōhoku Daigaku Bungakubu Kokushi Kenkyūjo), issue titled *Toyoda-Ishii ryōsensei zōkan kinengō* (Feb. 1973): 47–54.

Shioya Tsutomu. *Buwakebayashi seido no shiteki kenkyū; buwake-bayashi yori bunshūrin e no tenkai* [Historical study of the shared-forest system: Development from shared forests to divided forests]. Tokyo: Rin'ya kyōzaikai, 1959.

Suñaga Akira. "Kinugawa jōryūiki ni okeru ringyō chitai no keisei" [The formation of a timber area on the upper Kinugawa]. *Tochigi kenshi kenkyū* 212 (Dec. 1981): 1–32.

Takase Tamotsu. "Kaga han rinsei no seiritsu ni tsuite" [Regarding the establishment of a forestry system in Kaga]. *TRK Kenkyū kiyō* 54 (Mar. 1980): 519–534.

Toba Masao. *Nihon ringyōshi* [A history of Japanese forestry]. Tokyo: Yūzankaku, 1951.

Tokoro Mitsuo. "Ieyasu kurairichi jidai no Kiso kanjō shiryō" [Financial records of Kiso during its years as warehouse land of Ieyasu]. *TRK Kenkyū kiyō* 42 (Mar. 1968): 309–334.

———. *Kinsei ringyōshi no kenkyū* [Studies in the history of early modern forestry]. Tokyo: Yoshikawa kōbunkan, 1980.

———. "Nishikori tsunaba ni tsuite—kuchie kaisetsu" [The boom at Nishikori: An introductory explanation]. *Shakai keizai shigaku* 2, 12 (Mar. 1933): 101–118.

———. "Ringyō" [Forestry]. In *Taikei Nihonshi sōsho (sangyōshi)*, 198–223. Tokyo: Yamakawa shuppansha, 1965.

———. "Suminokura Yōichi to Kisoyama" [Suminokura Yōichi and the Kiso mountains]. *TRK Kenkyū kiyō* 43 (Mar. 1969): 1–26.

———. "Unzai chūkei kichi to shite no Inuyama" [Inuyama: Connecting point in lumber transport]. *TRK Kenkyū kiyō* 42 (Mar. 1968): 1–34.

Tokugawa Rinseishi Kenkyūjo, ed., *Tokugawa Rinseishi Kenkyūjo Kenkyū kiyō* [Bulletin of the Tokugawa Institute for the History of Forestry]. Tokyo: Tokugawa Reimeikai, annual. In 1989, with the start of the new Heisei year period, volume numbers changed from *nengō* of year preceding publication to number from date of first publication, so the 1988 volume (issued March 1989) is designated volume 23.

Tokugawa Yoshichika. *Kisoyama* [The Kiso mountains]. Tokyo: Dōrōsha, 1915.

Wakino Hiroshi. "Kinsei Nishikawa ringyō ni okeru zaimokushō keiei" [Merchant lumber management in the early modern Nishikawa area]. *TRK Kenkyū kiyō* 59 (Mar. 1985): 229–253.

Yamamoto Hikaru. "Azuchi Momoyama jidai no ringyō" [Forestry in the Azuchi-Momoyama period]. *RK* 226 (Aug. 1967): 20–28.

Yoshida Yoshiaki. *Kiba no rekishi* [A history of lumberyards]. Tokyo: Shinrin shigen sōgō taisaku kyōgikai, 1959.

INDEX

About the Author

CONRAD TOTMAN is professor of history at Yale
University. His personal experience with logging dates
from the era of axe, crosscut saw, and horse-drawn
sled, and he recalls fondly the smell of Eastern White
Pine in the morning. His acquaintance with Japan
began in 1954 and includes several years of residence
there. In 1981–1982 he enjoyed a year in Tokyo pursuing research at the Tokugawa Institute for the History
of Forestry, and he spent 1992–1993 in Kyoto teaching
at the Kyoto Center for Japanese Studies. He is the
author of several books on early modern Japan, including *The Origins of Japan's Modern Forests: The Case
of Akita* (Honolulu: University of Hawai'i Press, 1985);
The Green Archipelago, Forestry in Preindustrial Japan
(Berkeley: University of California Press, 1989); and
Early Modern Japan (Berkeley: University of California
Press, 1993).

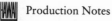 Production Notes

Composition and paging were done in
FrameMaker software on an AGFA AccuSet
Postscript Imagesetter by the design
and production staff of University of
Hawaii Press.

The text typeface is Sabon and the display
typeface is Trump.

Offset presswork and binding were done by
The Maple-Vail Book Manufacturing Group.
Text paper is Writers RR Offset, basis 50.